ISBN 978-3-540-02625-9   ISBN 978-3-540-37046-8 (eBook)
DOI 10.1007/978-3-540-37046-8

„Fortschritte der Hochpolymeren-Forschung/Advances in Polymer Science"

erscheinen zwanglos in einzeln berechneten Heften, die zu Bänden vereinigt werden.

Sie enthalten Fortschrittsberichte monographischen Charakters aus dem Gebiet der Physik und Chemie der Hochpolymeren mit ausführlichen Literaturzusammenstellungen. Sie sollen der Unterrichtung der auf diesen Gebieten Tätigen über solche Themen dienen, die in letzter Zeit besondere Aktualität gewonnen haben, bzw. die in neuerer Zeit eine lebhafte und nach literarischer Zusammenfassung verlangende Entwicklung erfahren haben.

Es ist ohne ausdrückliche Genehmigung des Verlages nicht gestattet, photographische Vervielfältigungen, Mikrofilme, Mikrophoto u. ä. von den Zeitschriftenheften, von einzelnen Beiträgen oder von Teilen daraus herzustellen.

Anschriften der Herausgeber:

Prof. Dr. J. D. Ferry, Department of Chemistry, The University of Wisconsin, Madison 6, Wisconsin/USA.

Prof. Dr. C. G. Overberger, Polytechnic Institute of Brooklyn, 333 Jay Street, Brooklyn 1, New York/USA.

Prof. Dr. G. V. Schulz, Institut für physikalische Chemie der Universität, Mainz.

Prof. Dr. A. J. Staverman, Fruinlaan 6, Leiden/Holland.

Prof. Dr. H. A. Stuart, Institut für physikalische Chemie der Universität, Mainz.

Springer-Verlag

Heidelberg	Berlin-Wilmersdorf
Neuenheimer Landstraße 28—30	Heidelberger Platz 3
Fernsprecher 279 01	Fernsprecher 830301
Fernschreiber 04-61723	Fernschreiber 01-83319

2. Band   Inhaltsverzeichnis   3. Heft

Seite

Bergsma, F., and Ch. A. Kruissink, Ion-Exchange Membranes. With 2 Figures 307

Porod, G., Anwendung und Ergebnisse der Röntgenkleinwinkelstreuung in festen Hochpolymeren. Mit 8 Abbildungen 363

Thomas, W. M., Mechanism of Acrylonitrile Polymerization. With 4 Figures 401

Sprung, M. M., Recent Progress in Silicone Chemistry. I. Hydrolysis of Reactive Silane Intermediates 442

Fortschr. Hochpolym.-Forsch., Bd. 2, S. 307—362 (1961)

Ion-Exchange Membranes

By

F. BERGSMA and CH. A. KRUISSINK

Central Laboratory T.N.O., Delft, Netherlands

With 2 Figures

Contents

Page

1. Introduction 307
2. Preparation of Ion-Exchange Membranes 310
3. Theory 314
   3.1. Irreversible Thermodynamics 314
   3.2. The Nernst-Planck Flux Equations 318
   3.3. Qualitative Considerations of the Membrane Behaviour on the Basis of M.S.T. Theory and Donnan Equilibrium 319
   3.4. Applications of the Different Flux Equations for the Calculation of Membrane Phenomena 322
   3.5. Important Quantities Connected with Electro Dialysis 338
   3.6. Application of the Absolute Reaction Rate Theory to Membrane Phenomena 342
4. Experimental Check of the Theoretical Equations 343
   4.1. Application of Irreversible Thermodynamics 343
   4.2. Application of the Nernst-Planck Flux Equations Combined with the M.S.T. Model 345
   4.3. Conductivity 352
5. Application of Ion-Exchange Membranes 354

Appendix 1. List of Symbols 357

Literature 357

1. Introduction

Ion-exchange membranes are a new development in the field of ion-selective membranes. By selectivity is understood the property that the transport numbers of the ions in the membranes have values different from those in the free solution. When the transport number of the cations is increased, the membrane is called cation-selective or negative, and the membrane is called anion-selective or positive in the opposite case.

The name "ion-exchange" membrane mainly refers to the fact that this ion-selective property can best be realized by using materials closely resembling the ion-exchange resins.

While the ion-exchange resins are applied in several different industrial processes in ever-increasing quantities, the ion-exchange membranes have found more or less distinct types of application only these last few years, e.g. in the desalting of brackish water. Scientific interest in selective membranes, however, dates already from the beginning of the present century when the occurrence of selectivity in biological systems was observed, e.g. in blood cells, frog skin, plant root cells etc.

In laboratory experiments, selective membranes were already applied years ago for the p_H-control during the electro-dialysis of p_H-sensitive colloids. In the three-compartment cells used, the electrode chambers were rinsed with distilled water. On account of the high mobility of the H^+ ions, the desalting cell was inclined to become acid. To oppose this effect, anode- and cathode membranes with different polarity were sought for. At the beginning the influence of these membranes on the current efficiency — i.e. the amount of salt removed per unit of charge flown through — was mentioned only sporadically (5, 23).

The investigations on membranes were hampered in the past by the lack of suitable membranes. The selectivity was low, especially at higher salt concentrations and the mechanical strength was in general insufficient for making reproducible measurements. As negative membranes were used e.g. porcelain plates, asbestos, sintered glass, or collodion and cellophane films. The choice of positive membranes especially was very limited. Examples are: woollen fabrics impregnated with gelatine, films dyed with basic dyes, animal membranes such as hog's bladder and leather. The disadvantage of gelatine and other animal membranes is that they are amphoteric: they can be used as positive membranes only in the p_H-region below their iso-electric points. As for dyed membranes, the dye is sometimes dissolved too easily from the film. As to the mechanism responsible for the ion-selective properties, it was understood at an early stage that the membrane structure had a large influence on its properties. Especially the charge of the membranes and their porosity were considered in order to explain the selectivity. Originally it was assumed that the charge occurred due to a difference in adsorption of positive and negative ions.

In 1935 T. Teorell (*168*) and in 1936 K. H. Meyer and J. F. Sievers (*97*) independently pictured a model for an ion-selective membrane which can account for its properties and the main features of which are therefore generally accepted now.

In the M.S.T.-model, the starting point is that fixed charged groups are connected to the membrane network e.g. in the case of negative membranes $-SO_3^-$ or $-COO^-$ groups and for positive membranes e.g. substituted ammonium groups, such as $-N(CH_3)_3^+$. The counterions of these groups can move freely in the water in the membrane. These counterions are the cause of it that in a positive membrane the share in the current transport of the anions is enlarged and that in a negative membrane the share of the cations is enlarged.

An essential point as to the M.S.T.-model is that it is assumed that the counterions are homogeneously distributed in the membrane-phase. MEYER and SIEVERS also report the ion-exchange properties of the membranes which they have made (98).

It was clear from the M.S.T.-model that the selectivity will be the better, the more charged groups have been incorporated in the material.

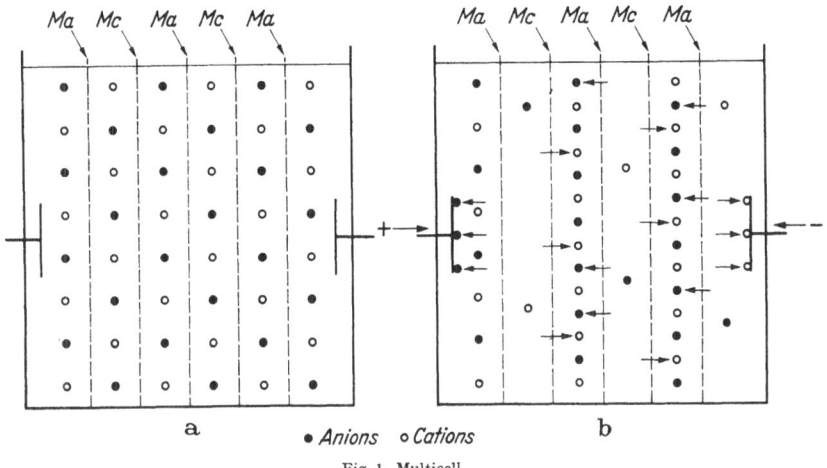

• Anions ○ Cations

Fig. 1. Multicell

Nevertheless it was not before 1949 that the most suited materials in this respect. viz. ion-exchange resins comparable with the technical ones, were successfully used in the preparation of the selective membranes (61, 68, 184).

One of the reasons is that initially it was very difficult to make ion-exchangers in the shape of films. An important stimulus for this research has been the application of the multicell for electrodialyses, which was already described by K. H. MEYER (99) in 1940 (see Fig. 1).

After 1950 the number of publications and patents relating to the making of selective membranes have steadily increased. Chapter 2 contains a survey of this development.

After the treatment of the theory and its experimental verifications in the chapters 3 and 4 respectively, chapter 5 will deal with the application of the selective membranes in electrodialysis and related fields.

2. Preparation of Ion-Exchange Membranes

The ion-exchange membranes consist of a macro-molecular network or coil, with attached to it the charged groups. The resin skeleton can be obtained through a polymerization reaction. Both the poly-condensation and the poly-addition can here be used. In the former case the macro-molecule arises because every time one component reacts with the chain, and a small molecule being split off. In the case of poly-addition a number of links are joined by a radical mechanism. Due to the introduction of the charged groups, swelling of the polymer occurs. When a very large number of groups is introduced, the linear polymer chains may even dissolve. In order to prevent this, cross-links are put between the chains. With the poly-condensation with a phenolic resin as a basis, these cross-links consist of $-CH_2$-bridges. These bridges are obtained by using a slight excess of formaldehyde over against phenol. The poly-addition of styrene occurs under addition of a bifunctional product, i.e. divinyl-benzene, so that the polystyrene network thus arising, possesses cross-links. With regard to the charged groups, following may be remarked: the negative membranes usually possess sulphonic acid groups. These are introduced by either sulphonating the monomer beforehand, or the synthetic resin afterwards.

In negative membranes also carboxylic acid groups occur.

These groups have weakly acid properties.

For the positive membranes the quarternary ammonium group is the most important one. It is mostly introduced by treating an aromatic nucleus with chloromethylether; in this way the $-CH_2Cl$-group is introduced. This group is made to react with a tertiary amine, and the quarternary ammonium compound thus arises. It is strongly basic. As other strongly basic groups, the pyridinium compounds are sometimes applied and, to a lesser extent, the sulphonium and arsonium groups. As less strongly basic groups, the tertiary, secondary and primary amines are used. It is perhaps useful to give some main values for ion-exchange membranes, by way of orientation.

The capacity — i.e. the number of charged groups per gram of dry matter — generally lies between 1 and 2 mgeq/gram dry. The water content may vary from 25 to 80% per gram dry; it is usually around 50%. The membrane thicknesses are generally between 100 and 500 μ. Under favourable conditions, the electrical resistance amounts to some Ohms for 1 cm² of a membrane. However, values of 20 Ohms cm² and more, occur too.

The first selective membranes based on commercial ion-exchangers were produced around 1950 (67, 68). Since that time about a hundred publications and patents concerning the preparation of ion-exchange membranes have appeared. In general, these membranes can be divided into five groups:

2.1. Homogeneous gel membranes.
    2.1.1. Polycondensation products.
    2.1.2. Polyaddition products.
2.2. Heterogeneous membranes of the ion-exchange resin plus binder type.
2.3. Chemically treated films.
    2.3.1. Starting-material hydrophylic films.
    2.3.2. Starting-material hydrophobic films.
2.4. Interpolymers.
2.5. Films on which polyelectrolytes are adsorbed.

2.1. Homogeneous Gel Membranes

2.1.1. Polycondensation products. This is the oldest type. During the preparation of ion-exchange resins a gel-stage often occurs. The condensation has then proceeded so far that, as it were, one large macromolecule has arisen which consists of linear chains that are inter-connected by cross-links. Generally these gels fall to pieces during swelling in water; however, if special precautions are taken, it is possible to keep them in the form of sheets, rods, etc.

Membranes of this type possess a high water content, a low concentration of active groups and a low electrical resistance. The selectivity is rather low at high solution concentrations. Also the mechanical properties are rather poor.

The base of the cation-exchange membrane is mostly phenolsulfonic acid. This is condensated with formaldehyde or paraform in an acid medium (61, 68, 116, 119, 133, 171). Sulfonated alkyl-aryl ethers are also used (32, 10). Instead of sulfonic-acid groups, sometimes carboxylic groups are introduced (10).

The anion-exchange membrane is prepared from alkylene poly amines likewise condensated with formaldehyde (7, 62).

To improve the mechanical strength, sometimes a backing is applied. Examples are glass fabric (6, 107) and parchment (136). The drawback is that the electrical resistance increases. Besides, on account of the difference in swelling between gel and backing microcracks can occur.

2.1.2. Polyaddition products. Ion-selective membranes are obtained by copolymerizing unsaturated compounds of which at least one posesses charged groups.

Examples are:

i. the copolymerizing of styrene, divinyl benzene and an ester or amide of a styrene sulfonic acid. After the reaction the membrane is hydrolyzed (42).

ii. Copolymers of vinyl aromatic compounds and N-alkyl-χ-vinyl pyridinium compounds (31).

iii. The preparation of anion-exchange resin membranes by reacting ethylenimine and epichlorohydrin (111).

iiii. Cation-exchange membranes with carboxylic acid groups are obtained by the copolymerizing of an olefinic carboxylic acid (or its anhydride, ester or acid chloride) and divinylbenzene (29, 175). Just as for the first-mentioned type, the mechanical properties are rather poor in general.

2.2. Heterogeneous Membranes of the Ion-Exchange Resin Plus Binder Type

A powdered ion-exchange resin is mixed with a thermoplast at a higher temperature. A rubber-mill can be used to that effect. Afterwards a membrane is formed by calandring or pressing. As the thermoplast is an electric insulator, it is obvious that the specific electric resistance of this type is rather high. This is so much the worse as the thickness of these membranes is rather high (minimum 0.3 mm) due to the way of their preparation. Therefore it is necessary to introduce as much ion-exchange resin as possible. However, the percentage of ion-exchanger has an upper limit on account of the impairment of the mechanical properties. On the average the content of ion-exchange resin amounts to 75 to 85%. Different preparation methods have been patented (17, 57, 186).

Polyethylene is often used as a binder. See e.g. (189).

Other binders are polyvinylchloride (112, 128), sometimes with a plasticizer (56); polyvinylacetal, specifically polyvinylformal plasticised with a copolymer of 1 : 3-butadiene and acrylonitrile (70); fluoro carbon polymers for preference a chlorotrifluoro ethylene polymer (132).

To improve the mechanical properties in some cases a backing is applied (183).

It is important that the ion-exchanger and the binder are very good mixed. To this end a solvent or swelling-agent for the binder is sometimes applied during mixing, which is removed before moulding (120). To improve the electrical resistance a binder with charged groups is used (170).

It is also possible to introduce afterwards in the resinous binder charged groups by a chemical treatment (187).

2.3. Chemically Treated Films

Starting with a film makes the membrane preparation rather easy. In general the apparatus can be used which is commonly applied for the treatment of films. In principle it is possible to make very thin membranes.

2.3.1. Starting-material hydrophylic films. On account of the hydro-phylic groups the water content of this type is usually high. This results in a low selectivity at high concentrations and a low electrical resistance.

As film material is used e.g. cellophane (16, 91), paper (135), and polyvinylalcohol (134, 101).

2.3.2. Starting-material hydrophobic films. This is a very promising type of ion-exchange membrane. In general the selectivity is high, in concentrated salt-solutions too. It is possible to combine a low electrical resistance with high mechanical strength.

Different kinds of films can be used as a basic material (149). In nearly all the cases cation-exchange membranes are obtained by sulfonating of films.

Examples are: the sulfonating of polyethylene film with chloro-sulfonic acid (60); the sulfonating of sheets of phenolformaldehyde resin (11); the treatment of a film consisting of polystyrene and poly-vinylchloride with concentrated sulfuric acid (4); the sulfonating of films consisting of aliphatic vinylpolymers with chlorosulfonic acid (125); the sulfonating of copolymers of a monovinyl- and a polyvinyl compound (30). Also are used copolymers of aromatic monovinyl-compounds and linear aliphatic polyene hydrocarbons (3); copolymers of an unsaturated aro-matic compound and an unsaturated aliphatic compound (76), and of reaction products of poly olefines and partially polymerized styrene (173).

The anion-exchange membranes usually are prepared by chloro-methylating of an aromatic nucleus, followed by aminating with e.g. trimethylamine, dimethylethanolamine, etc.

As films are used e.g. the polymerization product of ethylbenzene and divinylbenzene (33); the copolymer of styrene and butadiene (155); the copolymer of styrene and butadiene mixed with polyethylene (157); a vulcanized or cyclized copolymer of an aromatic vinylcompound and an aliphatic conjugated polyene (2). As a crack resisting matrix is mentioned the copolymer of styrene, divinylbenzene and butadiene with e.g. dioctylphthalate as a plasticizer (176). Other examples are the co-polymers of unsaturated aromatic compounds and unsaturated aliphatic compounds (77) and the reaction products of polyolefines and partially polymerized styrene (174). Primary groups can be introduced also with the help of Friedel-Crafts catalyst. Ts. KUWATA and co-workers treated a film of a copolymer of styrene and butadiene with an aluminium-ether complex and ethylenedichloride (79). Afterwards they allowed the film to react with trimethylamine. Another technique is the grafting of e.g. a polyethylene film with styrene (28).

Also it is possible to swell e.g. a polyethylene film in a mixture of styrene, divinylbenzene and a catalyst and after that to polymerize in the film (50, 58).

2.4. Interpolymers

This type of ion-exchange membrane is prepared by solving a polymer and a polyelectrolyte in one solvent. After that a film is casted and the solvent evaporated. For example a sulphonic interpolymer membrane can be prepared from a solution of linear polystyrene sulphonic acid, Dynel (a copolymer of acrylonitrile and vinylchloride) and dimethyl formamide (43). Other examples are given in (44, 45, 84, 169).

2.5. Films on which Polyelectrolytes are Adsorbed

Or this type only few examples are known.

So K. Sollner and collaborators developed the protamine collodion matrix membranes (81, 108, 162).

This type of membrane is rather rare, as two requirements must be satisfied: the amount of absorbed polyelectrolyte must be high to get a good selectivity, and the leaching-out must be negligible.

3. Theory

If a membrane separates two phases, a number of transport processes and related phenomena can take place depending on the external conditions, such as the concentrations of the ions in the surrounding solutions, the strength of an applied field, gradients of pressure and concentrations. To these phenomena which ultimately are due to migration of ions and solvent molecules the following belong: ion-transport, electro-endosmosis, diffusion and selfdiffusion, osmosis, hydraulic flow, streaming-current and potential, salt filtration, bi-ionic potentials and multi-ionic potentials. Aim of the theory is to correlate the various phenomena and to find the characteristic quantities, the knowledge of which is sufficient for the prediction of the behaviour of the membrane.

3.1. Irreversible Thermodynamics

The most general way to derive the relationships between the transport-phenomena is given by irreversible thermodynamics (46).

According to the principles hereof, the fluxes (Φ_i) of the components of the system are expressed as linear functions of properly defined "generalized" forces χ_i:

$$\Phi_i = \Sigma_k L_{ik} \chi_k .\tag{1}$$

The fluxes and forces must so be chosen that the entropy production per unit of time can be written as:

$$\frac{dS}{dt} = \Sigma \chi_i \Phi_i .\tag{2}$$

The L_{ik}'s are called the phenomenological coefficients, or "L-coefficients". They are independent of the process, but not necessarily independent of the concentrations.

The L_{ii}'s are a measure for the part of the flux Φ_i caused by direct action of the force χ_i on the species i. The L_{ik}'s $(i \neq k)$ are drag constants. These are due to the interaction of the species i with the species k.

ONSAGER's fundamental law states that:

$$L_{ik} = L_{ki} \tag{3}$$

Choosing for the fluxes the numbers of moles of each component passing per second through a cross-section of unit area, we get for the corresponding forces:

$$\chi_i = -\frac{1}{T} (z_i FE + v_i \Delta P + \Delta \mu_i) \tag{4}$$

z_i = charge of ion i
F = Faraday's constant
E = electrical potential difference
v_i = partial volume of ion i
ΔP = pressure difference
$\Delta \mu_i$ = difference in chemical potential of ion i per g mol.

Dependent on the model of membrane used, the systems of irreversible thermodynamics can be divided in two groups:

Discontinuous systems. The membrane is regarded as a surface of discontinuity, hindering the movement of the different ions and molecules. The driving forces are in this case the differences in electrical potential, pressure and chemical potential (*165, 166*) [see equation (4)].

To describe the membrane processes, certain phenomenological quantities are used, which are functions of the L_{ik}'s.

For example the electrical conductance per unit area of the membrane is the quotient of the electrical current density $\left(I = \sum_i z_i F \Phi_i\right)$ and the applied electrical potential difference when measured in an experiment in which there are no differences in pressure and concentrations $(\Delta \mu_i = 0)$ on both sides of the membrane. Making use of (1) and (4):

$$L_E = \left(\frac{I}{E}\right)_{\Delta P = 0, \, \Delta \mu = 0} \tag{5}$$

$$= \frac{1}{E} \cdot \sum_i z_i F \Phi_i \qquad (\Delta P = 0, \, \Delta \mu = 0)$$

$$= \sum_i \sum_k L_{ik} z_i z_k F^2 . \tag{6}$$

The factor $1/T$ is included in the L_{ik}'s.

The electrical transport number of ion i, i.e. the part of the charge carried by ion i is:

$$T_i = \left(\frac{z_i F \Phi_i}{\sum_i z_i F \Phi_i} \right)_{\Delta \mu = 0, \, \Delta P = 0} = \frac{z_i \sum_k L_{ik} z_k}{L_E} . \tag{7}$$

The reduced transport number or transference number — i.e. the number of moles of species i transported during the passage of one Faraday of charge is:

$$t_i = \frac{T_i}{z_i} = \frac{1}{L_E} \sum_k L_{ik} z_k . \tag{8}$$

From formula (8) it is seen that also uncharged molecules, e.g. the molecules of the solvent, may have a transference number with a positive value.

STAVERMAN (*165*, *166*) has shown how a number of transport phenomena across membranes, such as ion migration, hydraulic permeability, and membrane potentials are related to the L_{ik}'s. In general, if there are n components, $n(n-1)$ independent measurements are required to estimate all L_{ik}'s. So for a system consisting of 3 components (e.g. sodium ions, chlorine ions and water) 6 independent measurements are necessary.

It is shown by J. W. LORIMER, E. I. BOTERENBROOD and J. J. HERMANS (*84*) how these six independent relations can be obtained.

It must be borne in mind, however, that the implied condition is made that the L_{ik}'s have the same values for the different measurements. As, however, the L_{ik}'s are concentration dependent and the concentration-profiles in the membrane change their shape dependent on the kind of experiment, this condition is not satisfied in general.

Continuous systems. The membrane is considered as a quasi-homogeneous intermediate phase. The driving forces are the gradients of the electrical potential, the pressure and the chemical potential.

The equations now are valid locally. In cases when the gradients are zero (e.g. when the conductivity is determined) there is no difference in relation between the L_{ik}'s (which now apply locally) and the phenomenological coefficients according to the discontinuous approach. In case the gradients do not equal zero, the local equations yield diffusion equations which only after integration would yield the said phenomenological quantities. This integration can only be performed if the concentration dependence of the L_{ik}'s is known, which is generally not so. Therefore no examples of this treatment are known. However, this approach can be applied if drastic symplifications are introduced, viz., if the drag coefficients $[(L_{ik})_{i \neq k}]$ are assumed to be zero and if the remaining L_{ii}'s are expressed in the diffusion-coefficients which may then assumed to be still independent of the concentration.

This leads to the refined Nernst Planck equations (cf. § 2).

SPIEGLER (*164*) followed another way. Instead of the L_{ik}'s, he introduces the Ω_{ik}'s, which have the character of frictional coefficients and are defined by the phenomenological coefficients in the equations, which now express the forces as linear functions of the fluxes. It can be demonstrated that equations (1) may be written as:

$$F_{ik} = -c_i r_{ik}(\tilde{u}_i - \tilde{u}_k) = -c_i r_{ik} \tilde{u}_{ik} \qquad (9)$$

where:

c_i = concentration of component i

$\tilde{u}_i$ = velocity of component i

F_{ik} = friction force between components i and k

r_{ik} = "reduced friction coefficient" between components i and k.
      It refers to unit amounts of both kinds of particles.

Equation (9) states that the friction force F_{ik}, which the species i exert on the species k is proportional to the concentration of i and to the relative velocity of i and k. The proportionality factor is called the friction coefficient. The r_{ik}'s can be expressed in the L_{ik}'s. Also it can be shown that

$$r_{ik} = r_{ki} . \qquad (10)$$

The resultant of the friction forces exerted on i is equal to the generalized force χ_i.

Introduction of the friction model has two advantages:

1. One may hope that the r_{ik}'s are less concentration-dependent than the L_{ik}'s.

2. Calculations performed with r_{ik}'s are found to yield simpler and more straightforward forms than those with L_{ik}'s.

It is necessary to point out a difficulty with regard to the integration of the flux equations in a real membrane. If, for example, membranes with a pore structure are concerned, the final result which one calculates for a complicated network of capillaries which run in all directions and which are interconnected is different from what is calculated for the model which only contains pores which run perpendicularly to the membrane surface, but when proceeding from the local parameters (e.g. Ω_{ik}'s or diffusion coefficients) to the integral ones, an extra parameter occurs in the resulting expressions, which accounts for the nature of the pore structure (tortuosity factor).

SPIEGLER has used the friction model to describe a system consisting of sodium ions (1), chloride ions (2), water (3) and a charged matrix (4). He neglects the interaction of the sodium ions with the chloride ions. Then five independent measurements are needed to calculate the friction coefficients. SPIEGLER chose to be measured the self-diffusion coefficient of the sodium ions and the chloride ions, the specific electrical conductance, the transference number of one ion and the watertransport. The selfdiffusion coefficient of the sodium ions is:

$$\bar{D}_1 = \frac{RT}{\Omega_{13} + \Omega_{14}} \tag{11}$$

in which Ω_{ik} is defined as $\Omega_{ik} \equiv c_i r_{ik}$.

That of the chloride ions is:

$$\bar{D}_2 = \frac{RT}{\Omega_{23} + \Omega_{24}} \tag{12}$$

The specific electrical conductance is:

$$\varkappa = L_E \cdot d = F(\Phi_1 - \Phi_2)_{\left(\frac{dE}{dx} = 1,\ \Delta P = 0,\ \Delta\mu = 0\right)}$$
$$= \left(\frac{F^2}{a}\right) [c_1(\Omega_{23} + \Omega_{24})(c_1\Omega_{13} + c_3\Omega_{34}) + c_2(\Omega_{13} + \Omega_{14})(c_2\Omega_{23} + c_3\Omega_{34}) +$$
$$+ c_1 c_2(\Omega_{23}\Omega_{24} - 2\,\Omega_{13}\Omega_{23} + \Omega_{13}\Omega_{14})]\,, \tag{13}$$

in which d = membrane thickness, F = Faraday's constant, and a is defined as:

$$a = c_1\Omega_{13}\Omega_{14}(\Omega_{23} + \Omega_{24}) + c_2\Omega_{23}\Omega_{24}(\Omega_{13} + \Omega_{14}) +$$
$$+ c_3\Omega_{34}(\Omega_{13} + \Omega_{14})(\Omega_{23} + \Omega_{24})\,.$$

The transportnumber of the sodium ions is:

$$t^+ = \frac{\Phi_1}{\Phi_1 - \Phi_2} = \frac{F^2 c_1}{L_E \cdot d \cdot a} \times$$
$$\times [(c_1\Omega_{13} + c_3\Omega_{34})(\Omega_{23} + \Omega_{24}) + c_2(\Omega_{23}\Omega_{24} - \Omega_{13}\Omega_{23})]. \tag{14}$$

The electro-osmotic watertransport is:

$$\Phi_3 = \left(\frac{Fc_3}{a}\right) [c_1\Omega_{13}(\Omega_{23} + \Omega_{24}) - c_2\Omega_{23}(\Omega_{13} + \Omega_{14})]\,. \tag{15}$$

From these 5 equations Ω_{13}, Ω_{14}, Ω_{23}, Ω_{24} and Ω_{34} can be calculated.

In connection with the treatment by SPIEGLER two remarks must be made.

i. As usual, SPIEGLER in his treatment proceeded from the MST model.

ii. The relatively simple treatment according to SPIEGLER is only possible if no concentration gradients occur. If this is the case however, the r_{ik}'s must be averaged over the concentration. The fact that the concentration profiles are generally different for the various types of measurements, here constitues an extra difficulty.

3.2. The Nernst-Planck (110, 127) Flux Equations

In the above it is already said that due to drastic simplification of the flux equations, the Nernst-Planck equations arise. To show this it is necessary to first write equation (1) in a somewhat different form. For the purpose, the velocities of the several components are split up in a velocity of the particles relative to one another and a velocity of translation of the common centre of gravity. The result is:

$$\Phi_i = \Sigma_k L'_{ik} \chi_k + c_i v \tag{16}$$

where v = velocity of the centre of gravity

c_i = concentration of component i.

It should be remarked that it depends on the membrane model to what extent c_i is expressed. If the membrane is considered to be a homogeneous gel, c_i is the number of g mol of component i per unit of volume of the swollen resin. For the pore model, c_i is the number of gmol per unit of volume of pore liquid. In equation (16) now all the L'_{ik}'s with $i \neq k$ are assumed to be equal to zero.

The result is:

$$\Phi_i = L'_{ii} \chi_i + c_i v \tag{17}$$

For the continuous case, the forces become

$$\chi_i = -\frac{1}{T} (z_i F \operatorname{grad} \varphi + v_i \operatorname{grad} p + \operatorname{grad} \mu_i) \tag{18}$$

where φ = electrical potential

p = pressure.

If there is no potential gradient nor a pressure gradient across the membrane and if no electro-osmotic flow occurs, and when also the activity coefficients are ignored, (17) and (18) yield:

$$\Phi_i = -\frac{L'_{ii}}{T} (R T \operatorname{grad} c_i)$$

According to Fick's first law:

$$\Phi_i = - D_i \operatorname{grad} c_i$$

where D_i = diffusion coefficient of component i.

From this it follows that $D_i = L'_{ii}/T$.

Combining this with (17) and (18) we obtain:

$$\Phi_i = - D_i \left(\operatorname{grad} c_i + z_i c_i \frac{F}{RT} \operatorname{grad} \varphi + c_i \operatorname{grad} f_i \right) + c_i v \tag{19}$$

The equations (19) may be considered as refined Nernst-Planck equations. These equations combined with the M.S.T. model are treated by F. HELFFERICH (55) in a very exhaustive manner in the chapter on ion-exchange resin membranes of his book on ion-exchange resins. Extensive literature references are also given here.

3.3. Qualitative Considerations of the Membrane Behaviour on the Basis of M.S.T. Theory and Donnan Equilibrium

Previous to giving a quantitative elaboration of the Nernst-Planck equations for the different membrane processes, at first a qualitative treatment of membrane phenomena will be given here on the basis of M.S.T. theory and Donnan equilibrium.

An ion-exchange membrane consists of a continuous resin network and a continuous aqueous network which are interpenetrating. As to the membrane structure, the homogeneous gel and the pore structure should be distinguished. The homogeneous gel can perhaps best be described as a kind of "solid" solution of polyelectrolyte in water. It has been found, however, that the majority of homogeneous gel membranes have a micro structure (35, 147), i.e. they are inhomogeneous of structure, if considered in colloidal dimensions. In these instances, the pore model very probably gives a better description again.

To the resin network charged groups are connected, whose charges are neutralised by counterions (gegen ions), which can move freely in the water phase in the membrane. It is assumed that the distribution of resin network, fixed charges, solvent and ions is homogeneous. The M.S.T. model can be dirived from the older Helmholtz-Smoluchowski model. Here we have charged pores with a diffuse double layer (Gouy) near the wall. The solution in the inner part of the pores has the same composition as that of the outer solution. If the pore charge increases, the thickness of the double layer increases too and finally the double layer fills up homogeneously the whole pore. Then we have the M.S.T. model. See e.g. H. J. Oel (117) who shows very clearly that the M. S. T. model and the Helmholtz-Smoluchowsky model are two limiting cases of charge distribution in a membrane.

If an ion-exchange membrane is placed in a salt solution with the same counter-ion, some salt is absorbed and so in the membrane also some free ions occur with the same kind of charge as the fixed groups (co-ions).

When a uni-univalent salt is concerned, the following applies in equilibrium:

$$\frac{\bar{a}_+ \cdot \bar{a}_-}{a_+ \cdot a_-} = K \tag{20}$$

where $\bar{a}_+$ and $\bar{a}_-$ are the activities of the positive, respectively negative, ions in the membrane and a_+ and a_- those in the free solution. $K =$ constant of equilibrium.

If $\bar{f}_i \bar{c}_i$ is substituted for $\bar{a}_i$, it depends on the membrane model how c_i is expressed. For a pore model, c_i is the concentration of ion i in gequiv. per liter imbibed solution. For a homogeneous gel, c_i is the number of gequiv. of the ion i per unit of volume of the swollen resin. In the activity coefficient $\bar{f}_i$ must then be incorporated the interaction energy with the charged resin skeleton.

The value of K is given by:

$$-RT \ln K = \Delta \mu_0 \,,$$

where $\Delta \mu_0$ is the difference of the standard potentials in the membrane and in the free solution. If the same standard condition in both phases is proceeded from, then $\Delta \mu_0 = 0$ and $K = 1$. (20) then changes to the well-known Donnan relation:

$$a_+ \cdot a_- = \bar{a}_+ \cdot \bar{a}_- \tag{21}$$

If we neglect activity coefficients, then on account of electroneutrality:

$$c^2 = \bar{y}(\bar{y} + A) \tag{22}$$

where y = concentration of the co-ions in the membrane.

A = selectivity constant ≡ concentration of active groups per cm³ imbibed water.

In dilute salt solutions $c \ll A$; this gives:

$$y \ll c$$

Thus the number of counterions far exceeds the number of co-ions. In the boundary membrane-solution a potential jumb occurs due to the difference in concentration on both sides of this boundary plane. It is assumed that a state of equilibrium applies here, so:

$$F\varphi + \mu_i = F\bar{\varphi} + \bar{\mu}_i \tag{23}$$

because:

$$\mu_i = \mu_0 + RT \ln a_i$$

the following applies:

$$\bar{\varphi} - \varphi = \frac{RT}{F} \ln \frac{a_i}{\bar{a}_i}.$$

This potential difference is called Donnan potential and is given by:

$$E_D = \frac{RT}{F} \ln \frac{a_i}{\bar{a}_i}. \tag{24}$$

It is possible now to explain a number of important membrane properties in a qualitative way. The permselectivity i. e. the increased transference number of cations in a cation-exchange membrane respectively of anions in an anion-exchange membrane compared with the free solutions is due to the fact that the number of counterions is much higher than the number of co-ions.

The high permselectivity of ion-exchange resin membranes — due to the high A-values — gives rise to a high current efficiency when they are used in electro dialysis. This becomes clear when Fig. 1 is considered. Per unit of transported charge, a large quantity of salt is removed (maximum 1 gram-equivalent per Faraday of charge).

The high concentration of mobile ions in the membrane — also due to the high A-values — makes that in general their conductivity is high. This reduces the energy consumption of the electrodialysis.

At not too high concentrations of the outer solution, the amount of absorbed salt in the membrane in equilibrium is very low (y is very small). For this reason the diffusion of salt through a membrane is very small too. The membrane behaves as a barrier for salt diffusion. This is also favourable in electrodialysis, where high differences can occur in the salt concentrations of dialysate and concentrate. As the back-diffusion opposes the effect of the electrical desalting, its value must be as small as possible.

As the permeability of the membrane for ions of different charge signs largely varies, salt diffusion through a membrane is accompanied by the establishment of a membrane potential. These concentration or dialysis potentials play an important part in the study of membrane phenomena. With the above described model, the phenomenon of electro-endosmosis i. e. the transport of solvent across a membrane under the influence of an electric field, can easily be explained also.

As the number of counter-ions exceeds the number of co-ions by far, the total electrical force exerted on the counter-ions is higher than that exerted on the co-ions. This gives rise to a resulting friction force on the solvent in the same direction as the flow of the counter-ions.

On account of the Donnan exclusion, the concentration of absorbed salt in the membrane — also called diffusible salt — is very low. Therefore it is explicable that if a salt solution is pressed through an ion exchange membrane, under certain conditions salt filtration occurs.

As non-electrolytes are not submitted to the Donnan exclusion, the diffusion of non-electrolytes proceeds much faster than that of electrolytes. Therefore in principle it is possible to separate non-electrolytes from electrolytes with the aid of ion-exchange membranes.

3.4. Applications of the Different Flux Equations for the Calculation of Membrane Phenomena

For the calculation of membrane phenomena as diffusion through membranes, membrane potentials, electrical resistance, transference numbers during electrodialysis, concentration profiles in the membrane under different circumstances, the flux equations have to be solved with the appropriate boundary-conditions.

In general the mathematical problems are very complicated. In most cases simplifications have to be introduced to get useful expressions, as the neglection of solvent flow and the omission of gradients of activity coefficients.

Frequently the mobilities of the ions and the concentration of active groups are supposed to be constant across the whole membrane. Including all the above-mentioned simplifications, R. Schlögl (144) succeeded in performing the general integration of the Nernst-Planck equations for the quasi stationary state:

$$\Phi_i = -\overline{D}_i \left(\frac{d\bar{c}_i}{dx} + z_i c_i \frac{F}{RT} \frac{d\varphi}{dx} \right) \tag{25}$$

for mixtures of arbitrary electrolytes.

As auxiliary conditions occur:

$$\Sigma_i z_i \bar{c}_i + \omega \overline{A}' = 0 \quad \text{(condition of neutrality)} \tag{26}$$

where

z_i = charge of ion i

$\bar{c}_i$ = concentration of ion i in the membrane per unit volume of the swollen membrane.

ω = sign of charge of the fixed groups.

$\overline{A}'$ = concentration of fixed groups per unit volume of the swollen membrane.

and

$$\text{div. } \Phi_i = 0 \quad \text{(stationary state)} \tag{27}$$

The concentration of all ions in the two surface layers in the membrane are considered to be given. These are related to those of the outer solution by a set of Donnan relations analogous to equation (24). SCHLÖGL calculated the fluxes, the profiles of the concentrations in the membrane and the membrane potential.

In spite of all the simplifications, the obtained formulae are very complicated and the calculations rather tedious.

In the instance of simple systems more details can be involved. The best that can be done is to investigate for each system separately, how with as less neglections and approximations as possible, useful expressions can be obtained.

Examples of this are given in the following sections.

3.4.1. Diffusion Across Membranes

In this case a concentration gradient is present across the membrane without an applied electric field. Two cases may be distinguished: a) At both sides of the membrane, there are the same counter-ions (cf. 3.4.1.1.). b) At both sides of the membrane, there are different counter-ions (cf 3.4.1.2.).

3.4.1.1. Concentration gradient Across a Membrane. In the instance that a membrane separates two solutions of the same electrolyte, but with different concentrations F. HELFFERICH (ref. 55, page 319) calculated the ion-fluxes and the profiles of the internal concentrations, starting from the Nernst-Planck equations. Gradients of activity coefficients could be involved. However, convection (osmosis) had to be neglected.

The auxiliary conditions are: electroneutrality, no electric current flow and a quasi-stationary state.

The boundary conditions are given by the salt concentrations on either side of the membrane.

In order to solve the equations it is in general necessary to effect numerical integration.

If the fluxes are to be expressed in an explicit form, then simplifications and omissions are generally necessary.

J. S. Mackie and P. Meares (86) could largely avoid these, however, by introducing an empirical formula of the activity coefficients. For the purpose they carried out sorption measurements at ion-exchange membranes.

For a cation-exchange membrane and uni-univalent electrolytes a relation:

$$\ln \bar{f}_{\pm} = B \ln (\bar{c}_+ \bar{c}_-) + C ,$$

in which B and C are constants, could be found that fitted the experimental points. Differentiation of this equation to $\bar{c}_+$ and introducing the neutrality condition yields:

$$d \ln (\bar{f}_+) = \frac{B (2 \bar{c}_+ - \bar{A}')}{\bar{c}_+ (\bar{c}_+ - \bar{A}')} d \bar{c}_+ . \tag{28}$$

In an analogous manner they obtained for a cation-exchange membrane and divalent cations:

$$d \ln (\bar{f}_+) = H d \bar{c}_+ , \tag{29}$$

in which H is a constant, also estimated experimentally.

On application of equation (28), respectively (29), it was found that, as regards simplifications, it was necessary only to assume the concentration of the active groups and the coefficients of diffusion to be constant, in order to be able to solve the flux equations. The osmosis could be accounted for.

For a cation exchange resin and an uni-univalent salt their ultimate result is:

$$\Phi_+ = \frac{\bar{D}}{d} \left[(\bar{c}_+^\alpha - \bar{c}_+^\omega) + \frac{z_- \bar{A} (\bar{u} - \bar{v})}{(z_+ - z_-) (\bar{u} z_+ - \bar{v} z_-)} \ln S \right] +$$
$$+ \frac{B \bar{D}}{d} \left[2 (\bar{c}_+^\alpha - \bar{c}_+^\omega) - \frac{\bar{A}' (\bar{u} z_+ + \bar{v} z_-)}{z_+ (\bar{u} z_+ - \bar{v} z_-)} \ln S \right] + \tag{30}$$
$$- \left[\frac{\bar{D}}{d} (\bar{c}_+^\alpha - \bar{c}_+^\omega) - \frac{z_- v}{z_+} \left(\frac{\bar{c}_-^\alpha e^{v d/D} - c_-^\omega}{1 - e^{v d/D}} \right) \right]$$

with the abbreviation

$$\ln S = \ln \left(\frac{\bar{u} z_+^2 c_+^\alpha + \bar{v} z_-^2 c_-^\alpha}{\bar{u} z_+^2 c_+^\omega + \bar{v} z_-^2 c_-^\omega} \right) .$$

$\overline{D}$ = diffusion coefficient of the salt in the membrane.

$$= \frac{(z_+ - z_-)\, RT\, \overline{u} \cdot \overline{v}}{(\overline{u}\, z_+ - \overline{v}\, z_-)}$$

where:

B = the constant of formula (28)

v = rate of osmotic flow.

$\overline{u}$ and $\overline{v}$ can be determined with the aid of self diffusion measurements. $\overline{c}_+^{\alpha}$ and $\overline{c}_+^{\omega}$ can be found with sorption measurements, and v can be determined experimentally.

With the aid of (30), Φ_+ can be calculated. MACKIE and MEARES compared the values calculated with the experimental values (cf. chapter 4).

R. SCHLÖGL (145) specially studied the osmosis. He solved the flux equations for the case of uni-univalent electrolytes. It was necessary to omit activity coefficients and to assume the coefficient of diffusion, as well as the concentration of active groups across the membrane, to be constant. The final result has a very complicated form. Profiles of the internal ion-concentrations and of the potential and the pressure can be calculated. The theory gives an explanation of the phenomena of "anomalous osmosis" and of incongruent salt flow. The anomalous osmosis is due to the combined action of an osmotic pressure gradient in the membrane and an electric field, set up by the difference in mobility of counterions and co-ions. The osmotic pressure difference alone gives a normal positive osmosis, i. e. a flow of solvent in the direction of the more concentrated solution. If the counter-ions have a greater mobility than the co-ions the electric field acts in the same direction as the osmotic solvent flow (anomalous positive osmosis). If the co-ions have a higher mobility, the two forces counteract and it is possible that the solvent flow is in the direction of the dilute solution (anomalous negative osmosis). In the case of anomalous positive osmosis and a high water permeability of the membrane, it is possible that some salt will move in the direction of the concentrated solution (incongruent salt flow).

H. J. OEL (118) solved the flux equations, introducing many simplifications. He omitted the activity coefficients and assumed constant mobilities in the membrane and a constant concentration of active groups across the membrane. Osmosis was neglected.

The most important conclusions drawn from all the above-mentioned calculations are:

a) The velocity of the diffusion is governed by the diffusion coefficient of the co-ion.

b) A membrane with a high concentration of active groups, placed between two diluted solutions acts as a barrier for salt transport.

c) Under the simplifiing conditions assumed by Oel it is derived that the flow of salt across an ionselective membrane depends on the square power of the solution concentrations.

In this paragraph should also be mentioned the calculations of P. Meares and H. H. Ussing (94).

These investigators are greatly interested in biological membranes. One of their aims is to be able to differentiate between active and passive transport through a membrane. In the former case it is assumed that transport occurs under the influence of a metabolic process, and in the latter case under the influence of electrochemical forces as they have always been discussed here. These investigators proceed from the assumption that when calculated and measured fluxes agree, only passive transport is concerned. Meares and Ussing calculated the ratio of the fluxes in both directions through a membrane — the so-called influx and outflux — inter alia when there is a concentration gradient across the membrane. They use the Nernst-Planck equations including the activity-coefficients and osmotic flow. If an isotope a is present on one side of the membrane and an isotope b on the other, then in the case that the boundary-conditions are:

$$x = 0 \quad a_a = a_a^\alpha \text{ and } a_b = 0$$
$$x = d \quad a_a = 0 \quad \text{and } a_b = a_b^\omega,$$

their ultimate result is:

$$\ln\left(-\frac{\Phi_a}{\Phi_b}\right) = \ln\frac{c_a^\alpha}{c^\omega} + \ln\frac{f_a^\alpha}{f_b^\omega} - \frac{V\Delta P}{RT} - \frac{zF\Delta\varphi}{RT} + \frac{vd}{D} \tag{31}$$

where:

v = partial volume of the isotope.
D = diffusion coefficient of the isotope.
ν = rate of osmotic flow.

3.4.1.2. Interdiffusion. If a membrane separates two solutions with two different counter-ions, but the same co-ion, interdiffusion of ions takes place. The mathematical treatment of these bi-ionic systems is about the same as of the diffusion of an electrolyte across a membrane (52, 54).

Helfferich calculated the fluxes of ions and the concentration-profiles. Steady-state conditions are assumed.

The flux of co-ions and of solvent are neglected.

As boundary conditions it is assumed that in either surface layer of the membrane only the same counterions occur as are present in the respective adjacent solution. The most important conclusions to be drawn from the calculations are:

i. The interdiffusion flow is proportional to the total concentration in the membrane and thus about proportional to the concentration of the active groups, and inversely proportional to the membrane thickness.

ii. The shape of the concentration profiles is determined by the ratio of the diffusion coefficients of the counterions. The slower moving ion is enriched in the membrane.

If a membrane separates two solutions with mixtures of counter-ions — in which each counter-ion is present only on one side of the membrane — and the same co-ion, we meet with a so-called multi-ionic system. These are also treated by F. HELFFERICH (53, 55) (ref. 55, page 327). An explicit solution of the flux equations in this case is obtained if the flow of co-ions is neglected and if all the counter-ions possess the same valency. Gradients of activity coefficients in the membrane and convection are also neglected. Diffusion coefficients and concentration of active groups are considered to be constant. It is assumed that there is equilibrium between the salt solution and the membrane surface on either side of the membrane.

The boundary conditions are given by the activity of the ions in the solutions.

For the ratio of the fluxes of two counterions, which are present on the same side of the membrane is found:

$$\frac{\Phi_1}{\Phi_2} = \frac{\bar{D}_1 a_1 \bar{f}_2}{\bar{D}_2 a_2 \bar{f}_1}. \tag{32}$$

a_1 and a_2 are the respective activities of the counterions 1 and 2 in the solution and $\bar{f}_1$ and $\bar{f}_2$ the activity coefficients in the membrane. It can be concluded that other factors being the same, the flux of an ion is the higher, the greater its affinity to the membrane is. As the most serious inaccuracy in the calculations HELFFERICH considers the assumption of constant activity-coefficients across the membrane. Starting from more qualitative considerations and disregarding activity coefficients, R. NEIHOF and K. SOLLNER (109) arrive at the same equation for the ratio of the fluxes.

Influence of unstirred layers near the membrane. Near the membrane there exist unstirred layers which under unfavourable conditions can exert a considerable influence on the fluxes and the membrane potential too. F. HELFFERICH (51) has drawn the attention to this effect. The thickness of these layers depends on the rate of stirring. Under good stirring conditions the film-thickness amounts to 20 to $1 \cdot 10^{-3}$ cm. Under extreme conditions it can be reduced to 10^{-4} cm. It is not always possible to eliminate its influence (139). The transport in the films is diffusion-controlled. In some cases the effect of the films can be involved in the calculations. As an example the case of selfdiffusion is given here. A cation-exchange resin separates two solutions of identical chemical composition. The cations on either side are isotopes of the same element.

The flux of isotopes is (51):

$$\Phi = - \frac{\bar{D}_+ \bar{c}}{d\,(1 + 2\,\bar{D}\bar{c}\,\delta/Dcd)}\,, \tag{33}$$

where:

$\bar{D}_+ =$ diffusion coefficient of the ion.
$\bar{c}\;\; =$ total internal concentration of the ion.
$d\;\; =$ thickness of the membrane.
$c\;\; =$ external concentration
$\delta\;\; =$ thickness of the unstirred film.

If $\dfrac{Dcd}{\bar{D}\bar{c}\,\delta} \gg 2$, we have to do with gel-kinetics; the diffusion is controlled by the membrane itself.

If $\dfrac{Dcd}{\bar{D}\bar{c}\,\delta} \ll 2$, we have film-kinetics; the diffusion process is governed by the adhering films. Film-kinetics is favoured by a high ion-mobility in the membrane, a high concentration of active groups, a small thickness of the membrane, low solution concentrations and poor stirring conditions. The high rate of exchange of counter-ions in bi-ionic systems makes that these systems incline to film-control, especially with dilute solutions.

Also in electrodialysis unstirred layers play an important role chiefly during the desalting of dilute solutions.

If an electrical current is passed across a cation-exchange membrane, salt depletion takes place in the surface layer facing the anode. Supply of salt occurs by diffusion across the unstirred film. Maximum diffusion flow occurs if the salt concentration near the membrane equals zero.

Then, in the stationnary state (123, 124),

$$i_{lim} = t_+\,i_{lim} + \frac{FDc_e}{\delta} \tag{34}$$

where:

$c_e\;\; =$ concentration in the desalting cell.
$\delta\;\; =$ thickness of the diffusion layer.
$D\;\; =$ diffusion coefficient of the salt.

as $t_- = 1 - t_+$,

$$i_{lim} = \frac{FDc_e}{\delta t_-}\,. \tag{35}$$

3.4.2. Membrane Potentials

In general the diffusion of ions through a membrane is attended with the building-up of a potential jump across the membrane. The membrane potential is the potential difference which occurs between a point in the solution at the one side of the membrane and a point in the solution at the other side of the membrane; it can be measured with two equal reference electrodes, e.g. calomel electrodes.

The membrane potential can be derived from the flux equations. In the scheme of the irreversible thermodynamics of discontinuous systems this is done as follows:

During the measurement $\Delta P = 0$ and $I = 0$, i.e. $I = \Sigma z_i F \Phi_i = 0$ and thus with $\Phi_i = \Sigma_k L_{ik} z_k F E + \Sigma_k L_{ik} \Delta \mu_k$, together with the equations (6) and (8) one obtains:

$$I = \Sigma_i \Sigma_k L_{ik} z_i z_k F^2 E + \Sigma_i \Sigma_k L_{ik} z_i F \Delta \mu_k = 0$$

$$= L_E E + \frac{L_E}{F} \Sigma_k t_k \Delta \mu_k = 0 .$$

Identifying E with the membrane potential (E_M) the following expression is found:

$$E_M = - \frac{RT}{F} \Sigma_k t_k \Delta \ln a_k \tag{36}$$

where t_k = transference number of species k, i.e. the number of moles of k transported per Faraday of charge. The summation has to be extended to the uncharged species too.

In the scheme of the irreversible thermodynamics of continuous systems the following can be derived in an analogous manner.

$$\frac{d\varphi}{dx} = - \frac{RT}{F} \Sigma_k t_k \frac{d \ln a_k}{dx} . \tag{37}$$

The above equation can be integrated across the membrane between the surface-layers α and ω of membrane-solution. Thus the diffusion potential (E_D) in the membrane is found. The result is:

$$E_D = - \frac{RT}{F} \int_\alpha^\omega \Sigma_k \bar{t}_k \, d \ln \bar{a}_k \tag{38}$$

where the quantities with a bar again apply to the membrane phase.

In order to obtain the membrane potential E_M, the two Donnan potentials, which occur in the surface-layers membrane-solution, must be added to the diffusion potential.

$$E_M = E_D + \frac{RT}{F} \ln \frac{a_k'}{\bar{a}_k^\alpha} - \frac{RT}{F} \ln \frac{a_k''}{\bar{a}_k^\omega} ,$$

where

a_k' = activity of ion k in solution 1 at the α side of the membrane

a_k'' = activity of ion k in solution 2 at the ω side of the membrane.

Since $\Sigma t_k = 1$, we can formally write:

$$E_M = -\frac{RT}{F}\left[\int\limits_{\text{sol. 1}}^{\alpha} \Sigma_k \bar{t}_k^\alpha \, d \ln a_k + \int\limits_{\alpha}^{\omega} \Sigma_k \bar{t}_k \, d \ln \bar{a}_k + \int\limits_{\omega}^{\text{sol. 2}} \Sigma_k \bar{t}_k^\omega \, d \ln a_k\right]$$

$$E_M = -\frac{RT}{F}\int\limits_{\text{sol. 1}}^{\text{sol. 2}} \Sigma_k \bar{t}_k \, d \ln a_k. \tag{39}$$

When the t_k's are constant across the membrane, (39) is changed into (36). Equation (38) for the diffusion potential can also be obtained by a quasi-thermodynamic method.

In order to obtain the membrane potential, any potential differences present in adhering liquid films must be taken into account, such in addition to the Donnan potentials. It should be observed that the splitting up of the membrane potential in diffusion potential, phase-boundary potentials and film potentials has met with opposition (49, 121).

For the calculation of the membrane potential E_M with the aid of (39), the transference numbers and the activities of the ions in each place in the membrane must be known. However, in general this is not the case.

An expression for the diffusion potential can also be obtained by integration of the Nernst-Planck flux equations. This integration is generally very complicated, so that further approximations must be introduced.

R. Schlögl (144) obtained, through his general integration of the Nernst-Planck equations, also values for the diffusion potential. The approximations in the calculations are the same as those used for the fluxes (cf. § 3.4).

The calculations are cumbersome. In general no closed expression is obtained for the potential.

For simple systems it is better not to apply the general integration method, since then fewer approximations will generally suffice. The most important of these cases will be discussed below. One of these is the system in which the membrane separates two solutions of the same electrolyte, which have a different concentration. We shall deal with this in paragraph 3.4.2.1. Another interesting case is that in which the counter ions at both sides of the membrane are different, i.e. the bi-ionic and multi-ionic systems. Further details are given in 3.4.2.2. A potential may also arise due to flow of an electrolyte solution through a membrane (streaming-potential) cf. 3.4.2.3.

3.4.2.1. Concentration- or Dialysis Potential. The membrane separates two solutions of the same electrolyte, but with different concentrations.

J. W. Lorimer, E. I. Boterenbrood en J. J. Hermans (84) have derived a formula for the concentration potential in the case of univalent ions. These authors proceeded from equation (39), which they derived in a slightly different manner. After some re-arrangements they obtained:

$$E_M = -\frac{RT}{F}\left[\int_{\text{sol. 1}}^{\text{sol. 2}} \left\{(2\,\bar{t}_+ - 1) - \frac{2\,m\,M}{1000}\,\bar{t}_{H_2O}\right\} d\ln a_{NaCl}\right] \quad (40)$$

where:

m = molality of the solution

M = molecular weight of the solvent.

In this derivation use is made of the Gibbs-Duhem relation, which for the solutions outside the membrane yields:

$$-d\,\mu_{H_2O} = \frac{m\,M}{1000}\,d\,\mu_{NaCl}. \quad (41)$$

Formula (40) has been written in this form in order to enable experimental verification (cf. Chapter 4). The variation of $\bar{t}_+$ across the membrane was found from transference measurements, the membrane being placed in NaCl solutions whose activity was varied from that of solution 1 to that of solution 2.

Proceeding from equation (38) G. Scatchard (137) derived, in the case of external concentrations which are low compared with the concentrations of active groups, for a cation-exchange membrane and a uni-univalent electrolyte:

$$E_M = -\frac{RT}{F}\left[\ln\frac{a_A''}{a_A'} - \int_{\alpha}^{\omega} \frac{\bar{u}_x}{\bar{u}_A}\left(\frac{a_+}{\bar{f}_+\bar{A}}\right)^2 - \frac{\bar{u}_w\,\bar{f}_+}{\bar{u}_A\,\bar{f}_+}\left(\frac{a_+}{\bar{f}_+\bar{A}}\right) d\ln a_{\pm}^2\right] \quad (42)$$

where:

$\bar{u}_x$ = mobility of the co-ions in the membrane

$\bar{u}_A$ = mobility of the counter-ions in the membrane

$\bar{f}_+$ = average molal activity coefficient in the membrane

$\bar{A}$ = concentration of active groups per ml solution

$\bar{u}_w$ = "mobility" of the water in the membrane.

Neglecting osmotic flow, it is possible to integrate the Nernst-Planck equations including activity coefficients (55, 142) (ref. 55, p. 338), an expression being obtained for the diffusion potential. Adding the Donnan potentials, the result is:

$$E_M = -\frac{RT}{z_A F}\left[\ln\frac{a_A''}{a_A'} - (z_+ - z_-)\int_{\alpha}^{\omega} \bar{t}_x\,d\ln a_+\right] \quad (43)$$

where a_A' and a_A'' are the activities of the counter ion on either side of the membrane.

$\bar{t}_x$ = transference number of the co-ions.

Taking for the diffusion potential the Henderson diffusion potential and adding the two Donnan potentials one gets the well-known formula of K. H. Meyer (97, 98), J. F. Sievers (97, 98) and T. Teorell (168).

For an uni-univalent electrolyte and a cation-exchange membrane they obtained:

$$E_M = \frac{RT}{F}\left[Z \ln \frac{X'' + \bar{A}Z}{X' + \bar{A}Z} - \frac{1}{2}\ln \frac{(X' + \bar{A})\,(X'' - \bar{A})}{(X' - \bar{A})\,(X'' + \bar{A})}\right] \tag{44}$$

where:

$$Z = \frac{\bar{u}_+ - \bar{u}_-}{\bar{u}_+ + \bar{u}_-},$$

$$X = \sqrt{4c^2 + \bar{A}^2}$$

$\bar{A}$ = concentration of active groups in the membrane.

c = concentration of the solution.

The use of the Henderson diffusion potential involves that the mobilities $\bar{u}_+$ and $\bar{u}_-$ are supposed to be constant, and that ideal solutions are assumed. Water transport is neglected.

Different refinements have been introduced in the course of time. Meyer and Bernfeld (100) accounted for the specificity of the membranes and introduced the partition-coefficients l_c and l_a for cations and anions. Instead of $\bar{A}$ the term $\dfrac{\bar{A}}{\sqrt{l_c\,l_a}} = \dfrac{\bar{A}}{l}$ now occurs in the M.S.T. formula.

It is assumed that l is independent of the solution concentration. About the same way was followed by Manecke and Bonhoeffer (89), who introduced the average activity coefficient of the salt in the membrane.

$$\bar{a}^2 = \bar{f}_\pm^2\,\bar{c}_+ \cdot \bar{c}_-$$

Moreover they accounted for activity coefficients in the solution. This corresponds with substituting $\bar{A}$ in the M.S.T. formula by $\bar{f}_+\bar{A}$, while $X = \sqrt{4a^2 + (\bar{f}_\pm\,\bar{A})^2}$.

If the activities on both sides of the membrane do not differ greatly, the concentration gradients in the membrane are small, and average constant transference numbers may be used as a first approximation. With this assumption the membrane potential reduces to:

$$E_M = -\frac{RT}{F}\,(\bar{t}_+ - \bar{t}_-)\ln\frac{a''}{a'}. \tag{45}$$

This formula is frequently used to derive transference numbers from e.m.f. measurements.

It must be borne in mind, however, that results may be affected unfavourably, because water transport has been neglected. If the concentrations of the outer solutions are very small compared with the concentrations of the active groups, to co-ions in the membrane can be neglected. Then (43) reduces to:

$$E_M = -\frac{RT}{z_+ F} \ln \frac{a''_+}{a'_+}$$ (46)

for cation-exchange membranes

$$E_M = \frac{RT}{|z_-| F} \ln \frac{a''_-}{a'_-}$$ (47)

for anion-exchange membranes.

These Nernst equations give the maximum value for the concentration potential that can be reached.

3.4.2.2. Bi-Ionic Potentials (B.I.P's).

If a membrane separates two salt solutions with two different counterions, but the same co-ion, the corresponding membrane potential is called bi-ionic potential. For the calculation of the B.I.P. this is split in a diffusion potential and two Donnan potentials. The diffusion potential can be calculated by proceeding from equation (37).

G. SCATCHARD (137) obtained for the case of a cation-exchange membrane and uni-univalent ions:

$$E_{BIP} = -\frac{RT}{F} \left[\ln \frac{a'_2 \, \bar{u}^\omega_2 \, \bar{f}^\alpha_1}{a'_1 \, \bar{u}^\alpha_1 \, \bar{f}^\omega_2} - \int_\alpha^\omega \frac{\Sigma_i^+ \bar{m}_i \, \bar{u}_i \, d \ln \bar{u}_i / \bar{f}_i}{\Sigma_i^+ \bar{m}_i \, \bar{u}_i} + \right.$$

$$\left. + \int_\alpha^\omega \bar{t}_x \frac{\Sigma_i^+ \bar{m}_i \, \bar{u}_i \, d \ln \bar{a}_i \, \bar{a}_x}{\Sigma_i^+ \bar{m}_i \, \bar{u}_i} + \int_\alpha^\omega \bar{t}_w \, d \ln \bar{a}_w \right].$$ (48)

a'_1 and a'_2 are the respective activities of the counter-ion 1 in the solution on the α side of the membrane and the ion 2 in the solution on the ω side of the membrane.

$\bar{u}^\alpha_1$ and $\bar{f}^\alpha_1$ are the mobility and the activity coefficient of ion 1 in the membrane face near solution 1 on the α side of the membrane.

$\bar{u}^\omega_2$ and $\bar{f}^\omega_2$ are the corresponding quantities for ion 2 in the ω surface.

$\bar{m}_i$ mobility of ion i in the membrane

$\bar{t}_x$ = transference number of the common co-ion

$\bar{a}_x$ = activity coefficient of the co-ion in the membrane

$\bar{t}_w$ − transference number of water

$\bar{a}_w$ = activity coefficient of the water.

The first term in Scatchard's formula is the main term, the others are correction terms. The first of these is a correction for the change in mobilities and activity-coefficients in the membrane, the second accounts for the transport by the negative ion and the third is a correction for the water transport.

Neglecting the first correction term means that $\frac{\bar{u}_1}{\bar{f}_1}$ and $\frac{\bar{u}_2}{\bar{f}_2}$ are assumed to be constant across the whole membrane.

Then, instead of $\dfrac{\bar{u}_1^{\alpha}}{\bar{f}_1^{\alpha}}$ and $\dfrac{\bar{u}_2^{\omega}}{\bar{f}_2^{\omega}}$ we may write $\dfrac{\bar{u}_1}{\bar{f}_1}$ and $\dfrac{\bar{u}_2}{\bar{f}_2}$. In that case, the main term becomes:

$$- \frac{RT}{F} \ln \frac{a_2' \, \bar{u}_2 \, \bar{f}_1}{a_1' \, \bar{f}_2 \, \bar{u}_1} \tag{49}$$

and it is possible to obtain "transport ratios" $\dfrac{\bar{u}_2/\bar{f}_2}{\bar{u}_1/\bar{f}_1}$ from B.I.P. measurements.

F. Helfferich (52) integrated the flux equations:

and

$$\left. \begin{aligned} \varPhi_1 &= - \bar{D}_1 \left[\frac{d \bar{c}_1}{dx} + z_1 \bar{c}_1 \frac{F}{RT} \frac{d\varphi}{dx} + \bar{c}_1 \frac{d \ln \bar{f}_1}{dx} \right] \\ \varPhi_2 &= - \bar{D}_2 \left[\frac{d \bar{c}_2}{dx} + z_2 \bar{c}_2 \frac{F}{RT} \frac{d\varphi}{dx} + \bar{c}_2 \frac{d \ln \bar{f}_2}{dx} \right] \end{aligned} \right\} \tag{50}$$

with the auxiliary conditions of electro neutrality, no electrical current flow and a quasi-stationary state, to obtain the diffusion potential. The difference with Scatchard's treatment is that here electro-osmotic flow had to be neglected.

The Donnan potentials are added.

He obtained in the case of equal equivalent concentration on either side of the membrane.

$$E_{BIP} = \frac{RT}{F} \left[\frac{\bar{D}_1 - \bar{D}_2}{\bar{D}_1 z_1 - \bar{D}_2 z_2} \ln \frac{\bar{D}_1 z_1}{\bar{D}_2 z_2} + \frac{z_1 - z_2}{z_1 z_2} \ln \frac{\bar{A}'}{c} + \right.$$
$$\left. - \int_{\alpha}^{\omega} (\bar{t}_1 \, d \ln \bar{f}_1 + \bar{t}_2 \, d \ln \bar{f}_2) + \ln \frac{(\bar{f}_2^{\omega})^{1/z_2}}{(\bar{f}_1^{\alpha})^{1/z_1}} + \ln \frac{(f_1')^{1/z_1}}{(f_2')^{1/z_2}} \right], \tag{51}$$

where c is the equivalent concentration of the solutions, and f_1' and f_2' are the activity-coefficients of the ions 1 and 2 in the solution.

In the case of constant $\dfrac{\bar{D}_1}{\bar{D}_2}$, $\dfrac{\bar{f}_2^{z_1}}{\bar{f}_1^{z_2}}$ and constant concentration of the active groups, he obtained (ref. 55, page 343):

$$E_{BIP} = \frac{RT}{F} \left[\frac{\bar{D}_2 - \bar{D}_1}{\bar{D}_1 z_1 - \bar{D}_2 z_2} \ln \frac{\bar{D}_2 z_2}{\bar{D}_1 z_1} + \ln \frac{\bar{f}_2^{1/z_2}}{\bar{f}_1^{1/z_1}} + \right.$$
$$\left. + \frac{z_1 - z_2}{z_1 z_2} \ln \frac{\bar{A}'}{c_1'} + \frac{1}{z_2} \ln \frac{c_1'}{c_2'} + \ln \frac{f_1'^{1/z_1}}{f_2'^{1/z_2}} \right]. \tag{52}$$

In the case of $z_1 = z_2$, the formula reduces to:

$$E_{BIP} = \frac{RT}{zF} \ln \frac{\bar{D}_1 a_1' \bar{f}_2}{\bar{D}_2 a_2' \bar{f}_1}. \tag{53}$$

If one neglects activity-coefficients in the membrane, the formula of C. E. Marshall (93) is obtained. Marshall looked upon the B.I.P. as a liquid junction potential and applied Henderson's formula in a straight-forward manner. He incorrectly substituted activities for concentrations in this formula.

In the instance of univalent ions the flux equations (50) can be integrated without the assumption $\dfrac{\bar{D}_1}{\bar{D}_2}$ is constant (12), and it is obtained:

$$E_{BIP} = \frac{RT}{F}\left[\ln\frac{a_1'\,\bar{D}_1^\alpha\,\bar{f}_2^\omega}{a_2'\,\bar{D}_2^\omega\,\bar{f}_1^\alpha} + \int\limits_\alpha^\omega t_1\,d\ln\frac{\bar{D}_1}{\bar{f}_1} + \int\limits_\alpha^\omega t_2\,d\ln\frac{\bar{D}_2}{\bar{f}_2}\right]. \quad (54)$$

Apart from omission of water transport and transport by the co-ion, this formula is identical with SCATCHARD's.

3.4.2.3. Streaming Potential and Streaming Current. If a salt solution is pressed through an ion-selective membrane as a rule a potential across the membrane is established. This potential is called the streaming potential. The phenomenon is due to the excess of charge of the pore liquid, which compensates the charge of the fixed groups. When the liquid in the membrane starts moving, this excess of charge will be dragged along.

A potential is built up, which tends to counteract the flow of the counter ions and accelerates the flow of the co-ions, in such a manner that the solution streaming out is electro-neutral. The potential can be measured in the usual way by placing two reference electrodes in the solutions on either side of the membrane. If these electrodes are short-circuited, an electrical current flows through the connecting wire. This current is called the streaming current.

These two phenomena are not very important in the instance of ion-exchange membranes. The membranes which in actual practice are the most valuable ones have a high hydraulic resistance, so that the effects measured are small.

The two phenomena are furthermore not very suitable for the characterization of membranes, because structural changes may easily occur due to the high pressures which have to be applied.

It may therefore suffice to refer to the pertinent paragraph in "Ionen-austauscher" by F. HELFFERICH (ref. 55, p. 356).

3.4.3. Diffusion with Electric Current

In this case there is an electric field applied externally, in addition to a concentration gradient across the membrane.

Pertinent calculations have been carried out by R. SCHLÖGL and M. SCHÖDEL (146). They considered a cation exchange membrane and a uni-univalent electrolyte. A constant concentration of active groups was assumed to occur across the whole membrane, as well as constant activity and diffusion coefficients. For simplification of the calculations it was assumed that the diffusion coefficients of the cation and the anion have the same value. These authors proceeded from the Nernst-Planck flux equations in the following form:

$$\Phi_+ = -\bar{D}_+ \frac{d\bar{c}_+}{dx} - \bar{D}_+ \bar{c}_+ \frac{d\varphi}{dx} + \bar{c}_+ v \tag{55}$$

$$\Phi_- = -\bar{D}_- \frac{d\bar{c}_-}{dx} + \bar{D}_- \bar{c}_- \frac{d\varphi}{dx} + \bar{c}_- v \tag{56}$$

v is called the average flow velocity in the membrane. It was assumed that the electric field strength is much greater than the osmotic pressure gradient. The following substitution was then made:

$$v = -F\bar{A} \frac{\Delta\varphi}{d} \cdot \frac{1}{\varrho_0} = -F \frac{\bar{A}'}{w} \frac{\Delta\varphi}{d} \cdot \frac{1}{\varrho_0} = -\frac{F}{RT} \cdot \bar{D}_0 \frac{\Delta\varphi}{d} \tag{57}$$

w = relative volume of the pores
ϱ_0 = flow resistance of the membrane
$\bar{D}_0$ = "diffusion coefficient" of the solvent. It is defined as

$$\bar{D}_0 \equiv RT \frac{\bar{A}'}{w} \cdot \frac{1}{\varrho_0} . \tag{58}$$

The integration of equations (55) and (56) they performed in a very ingenious manner by introducing two parameters r and s, defined by:

$$\Phi_+ = v \frac{\bar{A}'}{2} (1 + r)(1 + s) \tag{59}$$

$$\Phi_- = -v \frac{\bar{A}'}{2} (1 - r)(1 - s) \tag{60}$$

Two equations for r and s are found by integrating the flux equations. The boundary conditions are given by the values of the concentrations of the ions in the surface layers α and ω.

The result is:

$$\frac{r}{r - \frac{\bar{D}}{\bar{D}_0}} \log \frac{\bar{c}^\omega/\bar{A}' - r}{\bar{c}^\alpha/\bar{A}' - r} = \frac{s}{s - \frac{\bar{D}}{\bar{D}_0}} \log \frac{\bar{c}^\omega/\bar{A}' - s}{\bar{c}^\alpha/\bar{A}' - s} \tag{61}$$

and

$$\frac{r(1 + rs)}{r - \bar{D}/\bar{D}_0} \log \frac{\bar{c}^\omega/\bar{A}' - r}{\bar{c}^\alpha/\bar{A}' - r} = \frac{d}{\bar{D}\bar{A}'F} I \tag{62}$$

where

$$\bar{c} = \bar{c}_+ + \bar{c}_-$$
$$\bar{D} = \bar{D}_+ = \bar{D}_-$$

and

$$I = F(\Phi_+ - \Phi_-) .$$

The membrane potential follows directly from the applied I.

When r and s have been calculated, the integral transference number can be determined too, because

$$\bar{t}_+ = \frac{\Phi_+}{\Phi_+ - \Phi_-} = \frac{(1 + r)(1 + s)}{2(1 + rs)} . \tag{63}$$

Besides the transference numbers SCHLÖGL and SCHÖDEL were able to calculate the concentration profiles in the membrane. If the solution concentrations are different on both sides of the membrane, these profiles change if the direction of the current is reversed. By consequence, the electrical resistance depends on the direction of the current. This is called the rectifier effect.

As is shown, the authors include in their derivations the influence of electro-osmotic flow. In an earlier publication R. SCHLÖGL (144) neglected this effect. Comparing the results, one has to conclude that electro-osmosis has a big influence on the concentration profiles.

In order to be able to distinguish between active and passive transport through biological membranes, P. MEARES and H. H. USSING (95) likewise made a study of the fluxes through a membrane under the influence of diffusion together with an electric current. They studied the influxes and the outfluxes of sodium- and chloride ions at a cation exchange resin membrane. They started from the Nernst-Planck flux equations of the type:

$$\Phi_i = -\bar{u}_i \bar{c}_i \left(RT \frac{d \ln \bar{c}_i}{dx} + RT \frac{d \ln \bar{f}_i}{dx} + z_i F \frac{d \Psi}{dx} \right) + \bar{c}_i \nu.$$

In the case that a membrane separates two solutions containing any number of solutes, but differing only in that isotope a in solution 1 is replaced by isotope b in solution 2, they derived:

$$\Phi_a = \frac{\left(z_a \bar{u}_a - \frac{\omega}{\varrho_0} \bar{A} \right) F \Delta \varphi \cdot c \exp. S}{d (1 - \exp. S)} \qquad (64)$$

$$-\Phi_b = \frac{\left(z_a \bar{u}_a - \frac{\omega \bar{A}}{\varrho_0} \right) F \Delta \varphi \cdot c}{d (1 - \exp. S)} \qquad (65)$$

where

$$S = \frac{F \left(\frac{\omega \bar{A}}{\varrho_0} - z_a \bar{u}_a \right) \Delta \varphi}{\bar{u}_a RT}$$

ϱ_0 = flow resistance of the membrane.
$\bar{A}$ = concentration of charged groups per cm^3 of imbibed water.
ω = sign of charge of the fixed groups.
c = $c_a + c_b$.

c_a and c_b are the concentrations of the isotopes a and b in either solution. From (64) and (65) it is obtained:

$$\ln \left(-\frac{\Phi_a}{\Phi_b} \right) = z_a F \frac{\Delta \varphi}{RT} + \frac{\nu d}{D_a} \qquad (66)$$

D_a = diffusion coefficient of the isotope.
ν = rate of osmotic flow.

The net flux is equal to:

$$\Phi_a + \Phi_b = -\left(z_a \bar{u}_a - \frac{\omega \bar{A}}{\varrho_0}\right) F \frac{\Delta \varphi}{d} c. \tag{67}$$

For the equivalent conductivity of a or b in the membrane is derived:

$$\lambda_a = \frac{z_a F^2 \left(z_a \bar{u}_a - \frac{\omega \bar{A}}{\varrho_0}\right)}{|z_a|}. \tag{68}$$

This substituted in (64) and (65) gives:

$$\Phi_a = \frac{|z_a| \lambda_a \Delta \varphi \cdot c \exp. S}{z_a F d (1 - \exp. S)} \tag{69}$$

and

$$-\Phi_b = \frac{|z_a| \lambda_a \Delta \varphi \cdot c}{z_a F d (1 - \exp. S)} \tag{70}$$

with

$$S = -\frac{\lambda_a |z_a| \Delta \varphi}{z_a F D_a}. $$

3.4.4. Salt Filtration

If a salt solution is pressed through an ion-exchange membrane, salt filtration can take place caused by the Donnan exclusion.

Starting from the Nernst-Planck flux equations Helfferich (ref. 55, page 359) obtained for the concentration ratio of the original solution and the solution pressed through, in the case of a cation-exchange membrane:

$$\left(\frac{c}{c_s}\right)_{c \ll A} = w \left(\frac{\bar{A}'}{c}\right)^{-z_-/z_+} \left(\frac{\bar{f}_+}{f_\pm}\right)^{(z_+ - z_-)/z_+} \cdot \frac{z_+ \bar{u}_+}{z_+ \bar{u}_+ - z_- \bar{u}_-} \tag{71}$$

c = bulk concentration of the solution on the influx side.
c_s = concentration of the outflux solution.

The equation is valid only in the first moment of the filtration, as the salt concentration in the unstirred surface layer on the ingoing side rapidly increases. For this reason a great part of the filtration effect is lost.

3.5. Important Quantities Connected with Electro Dialysis

3.5.1. Electrical Conductivity of Ion-Exchange Membranes

The specific electrical conductivity of an ion selective membrane is given by

$$\bar{\varkappa} = \Sigma_i z_i^2 F \bar{c}_i \bar{u}_i + \frac{\bar{A}'^2 F^2}{\varrho_0 w}. \tag{72}$$

The first term is the regular electrolytic term, the second the convection term. This convection term is derived with the aid of $\bar{\varkappa}_{conv.} = \bar{A}' F \bar{u}_0$, where $\bar{u}_0$ is the apparent mobility of the pore liquid. This latter factor is given by: $\bar{u}_0 = \dfrac{F\bar{A}'}{\varrho_0\,w}$, where $\varrho_0 =$ flow resistance of the membrane and $w =$ relative volume of the pores of the membrane.

For use in electro-dialysis cells the conductivity of the membranes must be high. This can be achieved by a high internal concentration of movable ions, thus by a high concentration of charged groups. Further the mobilities of the ions in the membrane must be high. So far, however, there is no general theory which enables to calculate the mobility in the membrane starting from the situation in solution. But a number of factors can be indicated, which affect the mobilities.

i. *The available area of the membrane.* A part of the volume of the membrane is occupied by the polymer net work. If it is assumed that the polymer is statistically divided, the available area is:

$$A_a = A_g(1 - V_p) \tag{73}$$

$A_g =$ geometrical area.
$V_p =$ volume fraction of polymer.

ii. *The tortuosity factor.* In order to cross the membrane the ions will have to travel round the polymer chains. By that the path length is increased. If the ratio of the real path length divided by the geometrical path length is called θ, the apparent mobility is

$$u = u_w/\theta^2 \tag{74}$$

and the tortuosity factor

$$\Omega = \frac{u}{u_\omega} = \frac{1}{\theta^2} \tag{75}$$

$u_\omega =$ mobility in absence of polymer net work.

For instance considering the membrane as a system of randomly distributed pores $\dfrac{1}{\theta^2} = \overline{\cos^2}\,\alpha = \dfrac{1}{3}$.

In that case

$$u = \frac{1}{3}\,u_\omega \,. \tag{76}$$

Thus $\Omega = \dfrac{1}{3}$.

Applying the lattice model for polymer solutions J. S. MACKIE and P. MEARES (87) derived:

$$\theta = \frac{1 + V_p}{1 - V_p} \tag{77}$$

and

$$\Omega = \frac{(1 - V_p)^2}{(1 + V_p)^2} \tag{78}$$

iii. *Specific Effects.* The factors mentioned at i and ii affect the mobilities of all ions in the same way. Therefore, if geometrical factors would be the only factors that influence the mobilities, one should expect the same ratio of $\dfrac{u}{u_\omega}$ for all ions.

Actually this is not found. In general it appears that if a membrane has a specific affinity for a definite ion, the apparent mobility is lowered. Thus a specific factor has to be introduced, which accounts for the inter action of the ions with the membrane substance.

iiii. *Influence of the Outer-Concentration.* From conductivity- and self-diffusion measurements it is concluded that the mobility of univalent counter-ions strongly increases with increasing concentration of the outer-solution, whereas the mobility of the co-ions slightly decreases (*85, 126, 143*).

The increase in mobility of the counter ions can amount to a factor of 2 to 3 changing the outer concentration from 0,001 to 0.1 M.

Two different explanations are given with regard to this phenomenon.

a) The counter ions move by jumping from one active group to the other. As the counter ions are strongly attracted by these fixed charges, the "probability of transition" is not very high and thus the mobility is low. At higher solution concentration some co-ions are present in the membrane and now jumping can take place via these co-ions, which act as a carrier for the counter ions.

This explanation is inter alia given by R. Schlögl (*143*) and M. A. Peterson and H. P. Gregor (*126*).

b) D. Mackay and P. Meares (*85*) apply the double-layer theory to explain the observed variation of the mobility of the counter ions. They pose that as the sorbed concentration increases, the space charge will be concentrated more and more into regions close to the matrix. This, by creating a more powerful screening of the fixed charges by a small number of counter ions, would reduce the overall interaction with the matrix per mole of counter ions present in the resin.

Although for highly selective membranes the first explanation seems to be the most plausible one, it is feasible that for membranes with a high water content and a rather low capacity (such as the homogeneous gel membranes with which Meares and co-workers experimented) the second explanation is the right one.

iiiii. *The drag Factor.* Apart from the tortuosity factor, it is obvious that the conductivity of the membrane increases if the water content decreases. However, at very low water contents it is possible that the pore diameter approaches that of the ions in the pore. Then the viscous drag on these ions will increase. The drag factor is calculated by Gregor and Jacobson (to be published).

iiiiii. *Remaining factors.* Besides the above-mentioned factors, also some attention is paid to the influence of the capacity of the ion exchange membrane (*22*), i.e. the number of mgaeq. of fixed charges per gram of dry resin, and the mutual influencing of the mobilities if more kinds of gegen ions are present (*161*).

3.5.2. Transference Numbers

These are very important quantities as they determine the current efficiency of electro dialysis, i.e. the number of gram equivalents of salt removed during passage of one Faraday of charge.

In an electrodialytic cell this current efficiency at the desalting of NaCl solutions is given by

$$q = t_{Na^+}{}^{(-)} - t_{Na^+}{}^{(+)}$$

where:

$t_{Na^+}{}^{(-)}$ = transference number of the Na^+ ion in the negative membrane,

$t_{Na^+}{}^{(+)}$ = transference number of the Na^+ ion in the positive membrane.

Because $t_{Na^+}{}^{(+)} + t_{Cl^-}{}^{(+)} = 1$, it follows that:

$$q = t_{Na^+}{}^{(-)} + t_{Cl^-}{}^{(+)} - 1 .$$

For the general case:

$$q = \sum_+ z_i t_i{}^{(-)} + \sum_- z_i t_i{}^{(+)} - 1 \qquad (79)$$

where $\sum_+$ means a summation referring positive ions.

At low solution concentrations and high selective membranes is $q \sim 1$, as

$$\sum_+ z_i t_i{}^{(-)} \sim 1 \quad \text{and} \quad \sum_- z_i t_i{}^{(+)} \sim 1 .$$

At higher solution concentrations their value decreases slightly due to Donnan sorption.

If more kinds of counter ions are present, it is important to know the relative transference of each species. This depends on the relative concentration of the species and on its mobility. The former factor is determined by the affinity of the ion exchange membrane for the pertinent species. It has been mentioned already that a higher affinity is attended with a lower mobility. In general it is found that the first-mentioned factor predominates, so that a higher affinity is coupled with a relatively higher share in the charge transport.

A higher affinity for given ions can be obtained by introducing specific fixed charges, or by incorporating in the resin skeleton certain chemical groups which interact with given counter ions. An example of the first type is the membrane with hexanitrodiphenyl amide groups, selective for potassium (159, 182). To the second type belongs the charge-transfer bonding in molecules combined in polymeric structures, investigated by W. SLOUGH (160).

3.5.3. Water Transport

This is also an important phenomenon in electro dialysis, as it diminishes the quantity of treated water. The direction of water transport is, actually in most cases from the desalting cells to the concentration cells.

In desalting seawater the losses can amount to 20 to 30% of the volume of the treated water. The electro-osmotic flow per unit of current is:

$$D_I = \left(\frac{V}{I}\right)_{\Delta P = 0} = \frac{F\bar{A}'}{\varrho_0 \bar{\varkappa}} \tag{80}$$

$\varrho_0 = $ flow resistance of the membrane. Because

$$V = \bar{t}_0 v_0 + \Sigma_i v_i \bar{t}_i = -\omega D_I F ,$$

where

$v_i = $ partial volume of ion i per mol.
$v_0 = $ partial volume of the solvent per mol.
$\bar{t}_i = $ transference number of ion i.
$\bar{t}_0 = $ transference number of the solvent.
$\omega = $ sign of charge of the fixed groups.

is:

$$\bar{t}_0 = -\frac{\omega D_I F + \Sigma_i v_i \bar{t}_i}{v_0} . \tag{81}$$

The only practical possibility for restricting the water transport during electro dialysis is that of using ion exchange membranes with a high flow resistance.

In principle ϱ_0 can be measured by determining the hydraulic permeability D_H.

$$D_H = \left(\frac{V}{\Delta P}\right)_{E = 0} \tag{82}$$

$$D_H = \frac{w}{\varrho_0 d}$$

where $w = $ relative volume of the pores of the membrane.

According to (82), D_H is the hydraulic permeability measured at solutions which have been electrically short-circuited. Generally

$$D_H'' = \left(\frac{V}{\Delta P}\right)_{I = 0} \tag{83}$$

is measured (148).

It can be easily derived that:

$$D_H'' = \frac{w}{\varrho_0 d}\left(1 - \frac{(F\bar{A}')^2}{\varrho_0 \bar{\varkappa} w}\right). \tag{84}$$

3.6. Application of the Absolute Reaction Rate Theory to Membrane Phenomena

Another approach that can be made to the study of membrane phenomena is the application of the transition state theory (40).

J. F. Danielli (36) was the first to consider the membrane as a succession of potential barriers. Migration takes place in such a way that the elementary particles jump over the subsequent barriers.

According to the theory of absolute rate processes, the number of times per second a molecule jumps is given by:

$$k' = K \frac{kT}{h} e^{-\Delta F/RT} \tag{85}$$

K is the transmission coefficient, which is the fraction of the ions that, having reached the top of the barrier, proceed directly to the final state without returning to the initial state.

In most cases it is equal to unity.

$\frac{kT}{h}$ is a frequency factor involving the Boltzmann constant k, the absolute temperature T and the Planck constant h.

ΔF is the difference in free energy of the top and the foot of the barrier.

The flux is determined by the difference in the free energies of activation for the different unit processes. B. J. ZWOLINSKI, H. EYRING and C. E. REESE (190) have extended the approach of DANIELLI. Their derivations are only valid in the case that the energy barriers are regularly divided, and if the distance between these barriers is very small. These authors consider also the effect of external forces. They obtain if there is an electrical potential gradient and a concentration gradient across the membrane:

$$\Phi_i = -D_i \left(\frac{z_i F c_i}{RT} \frac{d\varphi}{dx} + \frac{c_i \, d \ln a_i}{dx} \right) \tag{86}$$

and

$$D_i = k' \lambda^2 \tag{87}$$

where

λ = distance between the barriers.

z_i = charge of ion i.

Thus, essentially, the Nernst-Planck flux equation is found. In principle, however, the theory enables the determination of the absolute values of the diffusion coefficients.

4. Experimental Check of the Theoretical Equations

In this chapter we will trace in how far the theoretical formulae, which for the greater part are derived under simplifying assumptions and sometimes by neglecting certain effects, are experimentally confirmed.

In general it can be said that the experimental material is not extensive. The experimental material concerns the application of irreversible thermodynamics, the application of refined Nernst-Planck flux equations and the application of quasi-thermodynamics. The latter is used to derive equations for membrane potentials.

4.1. Application of Irreversible Thermodynamics

The experimental material concerning direct application of irreversible thermodynamics is very scarce. This is partly due to the fact that it is a recent development.

J. W. Lorimer, E. I. Boterenbrood and J. J. Hermans (*84*) verified equation (40) by direct measurement of the transference numbers of the ions and of water at different concentrations of the outer solution.

Their membrane consisted of sodium carboxy methylcellulose incorporated in viscose.

Their membrane showed very high water transportnumbers viz. 100 to 300, depending on the outer concentration. They found a fair agreement between the experimental and theoretical membrane potentials. Further it was concluded that the watertransport contribution to the membrane potential is not negligible. It can be as high as 4%.

K. S. Spiegler (*164*) has shown the way in which it is possible to calculate friction coefficients from a series of different transport phenomena. For the system membrane-sodiumchloride solution he determined the selfdiffusion coefficient of sodium ions and chloride ions, the electroosmotic watertransport number, the specific conductance of the membrane and the transportnumber of the sodium ions in the membrane. Neglecting the interaction between counter ions and co-ions in the membrane he could calculate the friction-coefficients Ω_{13}, Ω_{14}, Ω_{23}, Ω_{24} and Ω_{34} with the aid of formulae (11)—(15).

Spiegler himself gives only qualitative considerations, concerning the usefulness of his theory.

So he calculated on the basis of experiments of Stewart and Graydon (*167*) the ratio of "reduced friction coefficients" $\frac{\Omega_{34}}{\Omega_{13}}$ for membranes with different watertransport numbers.

Ω_{34} is a measure for the friction between solvent and membrane matrix; Ω_{13} is determined by the friction between the Na^+ ions and the solvent.

Spiegler showed that while the watertransport number varied by a factor of about 4, the value of $\frac{\Omega_{34}}{\Omega_{13}}$ varied only over a range of $\pm 25\%$.

The way pointed out by Spiegler was followed by P. Meares (*96*) in his study of fluxes of sodium and chlorine ions across a cation-exchange resin membrane, separating two solutions of equal concentration of sodium chloride during the passage of an electric current.

Meares and his collaborators are especially interested in transport processes across biological membranes. They wish to distinguish experimentally between the active and the passive transport of a solute. For that purpose they determined the fluxes of the sodium ions in each direction through the membrane, using the technique of radio-tracers. The ratio of these experimental fluxes was compared with the theoretical ratios. The same is done with regard to the chlorine ions.

The theoretical fluxes were calculated with the aid of Ω_{ik}'s, which were derived from previous measurements such as proposed by SPIEGLER.

For the counter-ions a good agreement was found between the theoretical and the experimental flux ratios. However, the agreement was poor for the co-ions. It is suggested that this is due at least in part to neglecting the frictional interaction between co-ions and counter-ions. Therefore it is necessary that all the frictional coefficients are determined.

As long as this has not been done, the author (96) prefers, as regards the co-ions, the theoretical flux ratios based on the Nernst-Planck equations.

4.2. Application of the Nernst-Planck Flux Equations Combined with the M.S.T.-Model

In this field a fairly great deal of experimental work is done. In general it can be said that calculations on transport phenomena performed on this basis agree rather well with the experimental data. In many cases inclusion of a convection term is of importance.

4.2.1. Diffusion Across Membrane

4.2.1.1. Concentration Gradient Across a Membrane. J. S. MACKIE and P. MEARES (88) tested the equation (30), derived by them. They measured the diffusion of five different salts through a homogeneous gel membrane of the sulphonated phenolformaldehyde type. The osmotic flow during diffusion was measured too. For the mobilities in the resin the values in free solution were taken, multiplied with the factors Aa [equation (73)] and Ω [equation (78)].

In general the calculated fluxes appeared to be somewhat larger than the observed fluxes. The discrepancies were greatest at low solution concentrations. The authors attributed this to the immobilization at low concentration of the counter-ions by the electric field of the fixed charges. The deviations at high concentrations are ascribed to the fact that the electrophoretic effect is neglected. Also the effect of osmotic deswelling at higher solution concentration should be accounted for. The osmotic term tends to over-correct for high rates of solvent flow. According to the authors this is due to the assumptions made in its derivation.

As the mobilities in the membrane are calculated according to a crude statistical treatment, the authors conclude that the agreement between general magnitudes in theory and experiments is well within the limits of accuracy of the treatment.

Equation (31), giving the ratio of the fluxes of co-ions in both directions through the membrane was tested by P. MEARES and H. H. USSING (94). They measured the fluxes of sodium ions in both directions through a Zeo-Karb 315 membrane, separating two sodiumchloride solutions of different concentrations.

The same was done for the Cl⁻ ions. For this purpose the Na⁺ ions, respectively the Cl⁻ ions, were labelled in the concentrated solution or in the diluted NaCl solution.

Also the net fluxes were determined and compared with the theoretical values, obtained from equation (30).

A fairly good agreement was found between the observed and the calculated flux ratios, the former usually being the larger. It appeared that omission of the term

$$\frac{\bar{f}_a^\alpha}{\bar{f}_b^\omega}$$

from (31) gave a better agreement, even within the experimental accuracy (!). The authors were not able to draw any definite conclusions from this observation.

The theoretical net fluxes were found to be about 9% larger than the experimental ones.

4.2.1.2. Interdiffusion. F. Helfferich and H. D. Ocker (54) studied the interdiffusion of counterions through an ion-exchange membrane. Bases for their calculations were the Nernst-Planck flux equations combined with the M.S.T. model. Ion-fluxes and concentration profiles in the membrane were calculated.

The measurements were made at a homogeneous gel membrane, a condensation product of phenolsulfonic-acid and formaldehyde, with the ion pairs Na^+/H^+, Na^+/K^+ and Na^+/Sr^{++}.

The concentration profiles were measured on a stack of six membranes, separating the two solutions.

The experimental results proved the soundness of the theoretical basis.

4.2.2. Membrane Potentials

Membrane potentials are measured by placing two reference electrodes in the two solutions at both sides of the membrane. These reference electrodes may cause difficulties. This problem has been studied by M. Kahlweit (63).

In the simple case of a 100% selective membrane and AgCl reference electrodes, the following is found for the e.m.f. of the cell, as a sum of the membrane potential [equation (46)] and the two electrode potentials:

$$E = \frac{2\,RT}{F} \ln \frac{c^\omega f_\pm^\omega}{c^\alpha f_\pm^\alpha}. \tag{88}$$

When using reversible electrodes there are no difficulties. However, when calomel electrodes are used, the activity coefficients are found to be very complicated functions of concentrations and mobilities.

On the other hand, if the diffusion potential of the KCl bridge is considered to be zero, the activities of the single ions occur in the formula of the membrane potential. KAHLWEIT investigated to what extent corresponding values of membrane potentials were measured when AgCl and calomel electrodes were used. He applied a homogeneous gel membrane of the phenol sulphonic acid type. For HCl and KCl the cells with AgCl and calomel electrodes yielded equal values to respectively 0.2 n for HCl and 0.1 n for KCl. These values also agreed with formula (88), when values from the literature were substituted for the activity coefficients. However, for multivalent counter-ions and also for mixtures of electrolytes, considerable differences were found for both types of electrodes.

4.2.2.1. Concentration- or Dialysis Potential. The refined formula of MEYER, SIEVERS and TEORELL for the concentration potential has been experimentally tested by several investigators. G. MANECKE and K. F. BONHOEFFER (89) and K. F. BONHOEFFER and U. SCHINDEWOLF (21) found that for external concentrations smaller than appr. $^1/_3$ of the concentration at fixed charged groups the general trend of the potential with the external concentration was as indicated by the formula of K. H. MEYER and J. F. SIEVERS (97) [equation (44)]. Above this value there are deviations.

With the aid of equation (44) $\bar{f}_+ \bar{A}$ can be determined graphically; it is the product of the main activity of the mobile ions diffused in the membrane and the concentration of the charged groups. $\bar{f}_+$ can be found by determining $\bar{A}$ analytically. In this manner G. MANECKE (90) determined $\bar{f}_+$ for NaCl and KCl in a cation exchange membrane of the phenol sulphonic acid type. The relation $\dfrac{\bar{f}_{NaCl}}{\bar{f}_{KCl}}$ can also be found by measuring the equilibrium constant for the membrane in equilibrium with a solution of a mixture of NaCl and KCl.

The two methods yield corresponding values.

K. F. BONHOEFFER, L. MILLER and U. SCHINDEWOLF (20) performed similar measurements i.a. at cation exchange membranes with weakly acid groups and strongly acid groups and K^+, Na^+, Li^+ and H^+ ions.

It was found that for membranes with weakly acid, respectively weakly basic, groups there were no corresponding values for the two methods.

In the case of 100% selective membrane and negligible water transport, the membrane potential is given by the Nernst equations (46), respectively (47).

With the aid of these equations it can thus be investigated whether a membrane is indeed ideally permselective.

Also for the use of a membrane as membrane electrode it is necessary that equations (46) and (47) are satisfied.

In general it may be said that for the normal ion exchange membranes there is a concentration area for which the Nernst equation holds true. At high electrolyte concentrations deviations occur due to the participation of the co-ions in the transport. At very low concentrations deviations are found too. As causes are here mentioned: the influence of unstirred surface layers and hydrolysis of the charged groups. With membranes of a high permeability to the solvent, the Nernst potential is not reached, because the transport of the solvent reduces the membrane potential. The decrease is some per cent at most.

A number of examples of the testing of the Nernst equation follows now.

G. SCATCHARD (138) found that the requirements of the Nernst equations are met in the case of an anion- and cation selective membrane of the heterogeneous type in combination with HCl, NaCl and $CaCl_2$ solutions in the concentration range of approx. 10^{-4} n to approx. 0.1 n.

T. R. E. KRESSMAN (69) performed similar tests at a homogeneous gel membrane of the phenol sulfonic acid type and KCl solutions. In the range of $a_+ = 10^{-3.3}$ and $a_+ = 0.1$ the slope was found to be approx. equal to the theoretical slope, which corresponds with a potential change of 58 mV at a change in activity with a factor of 10. U. SCHINDEWOLF and K. F. BONHOEFFER (141), who likewise investigated a phenol sulfonic acid membrane, found results analogous to those of KRESSMAN.

W. F. GRAYDON and R. J. STEWART (41) also compared the membrane potentials with the values according to equation (46). The membrane investigated was a copolymer of p-styrene sulfonic acid and styrene crosslinked with divinyl benzene. In the large majority of cases the experimental values were lower than those according to equation (46). The smaller part of this difference could be attributed to the transport of the co-ions and was calculated roughly. The greater part was attributed to water transport. From this the transport number of water was calculated; it varied from 1 to about 60. It was found that the water transport was proportional to the water content and inversely proportional to the number of crosslinks. A provisional direct measurement was effected of a water transport number. The value corresponded rather well with the indirect determination as described above.

In a later article (167) are communicated the results pertaining to a larger number of water transport measurements.

It appeared that only for the membranes with very high selectivity the correction for water transport is sufficient to explain the difference between the ideal membrane potential according to (46) and the value measured. If the selectivity is less than 98%, a difference remains, also after correction of the water transport. The authors attribute this to the influence of unstirred surface layers.

Measurements at heterogeneous membranes, the potentials measured being compared with the Nernst potentials, have been carried out by A. G. WINGER, G. W. BODAMER and R. KUNIN (181) and by M. R. J. WYLLIE and H. W. PATNODE (184).

4.2.2.2. Bi- and Multi-Ionic Potential. When correction terms are neglected, the formula for the bi-ionic potential reads:

$$E_{BIP} = -\frac{RT}{F} \ln \frac{a_2 \bar{u}_2 \bar{f}_1}{a_1 \bar{u}_1 \bar{f}_2}. \tag{49}$$

The corresponding formula for the multi-ionic potential becomes:

$$E_{BIP} = -\frac{RT}{F} \ln \left(\Sigma_i^+ a_i \frac{\bar{u}_i}{\bar{f}_i} \right)_\omega \Big/ \left(\Sigma_i^+ a_i \frac{\bar{u}_i}{\bar{f}_i} \right)_\alpha. \tag{89}$$

Formulae (49) and (89) are obtained when it is assumed that the mobilities and the activity coefficients of all the counter-ions in the membrane remain constant across the whole thickness of the membrane.

This condition is certainly not fulfilled, because the composition of the membrane, as regards the counter-ions, is very much changed when one proceeds from the one boundary surface to the other. Particularly when the various counter-ions have greatly different affinities to the membrane, deviations will occur. In practice these are the most interesting cases, because the B.I.P.'s are used for the study of this specific selective behaviour of the membranes. Assuming that formula (49) is correct, "transport ratios" $\frac{\bar{u}_1}{\bar{f}_1} \Big/ \frac{\bar{u}_2}{\bar{f}_2}$ are determined with the aid of B.I.P.'s.

Formulae (49) and (89) have been experimentally tested by some investigators. R. NEIHOF and K. SOLLNER (109) compared the "transport ratios" obtained from B.I.P.'s with those obtained from diffusion measurements. In their experiments two counter-ions 1 and 2 occurred at the α side of the membrane, and at the ω side there was a third counter-ion. In this case,

$$\frac{\Phi_1}{\Phi_2} = \frac{a_1 \bar{u}_1 \bar{f}_2}{a_2 \bar{f}_1 \bar{u}_2} \tag{90}$$

as was derived by F. HELFFERICH and R. SCHLÖGL (53). By measuring $\frac{\Phi_1}{\Phi_2}$ at given a_1 and a_2 they also found $\frac{\bar{u}_1}{\bar{f}_1} \Big/ \frac{\bar{u}_2}{\bar{f}_2}$.

As a negative membrane was used a collodion membrane at which sulfonated polystyrene had adsorbed. As counter-ions K^+, Na^+, Li^+ and H^+ were chosen.

Also the positive membranes were of the collodion type at which poly-2-vinyl-N-methyl pyridinium chloride or protamine was, adsorbed.

Here the counter-ions were: Cl^-, acetate, iodate, nitrate, bromide, iodide and rhodanide.

In many cases the agreement between the transport ratios obtained in the two different manners is reasonably good.

However, larger differences may occur too.

Another test which was applied by K. Sollner and co-workers (38, 163) consists in comparing the measured multi-ionic potential with the calculated one according to formula (89), in which are substituted the "transport ratios" determined from the corresponding B.I.P.'s. For example, in the instance that three different counter-ions occur on either side of the membrane, the following is found with the aid of equation (89):

$$E_{BIP} = -\frac{RT}{F} \ln \frac{a_1^\omega + a_2^\omega \frac{\bar{u}_2}{\bar{f}_2}\Big/\frac{\bar{u}_1}{\bar{f}_1} + a_3^\omega \frac{\bar{u}_3}{\bar{f}_3}\Big/\frac{\bar{u}_1}{\bar{f}_1}}{a_1^\alpha + a_2^\alpha \frac{\bar{u}_2}{\bar{f}_2}\Big/\frac{\bar{u}_1}{\bar{f}_1} + a_3^\alpha \frac{\bar{u}_3}{\bar{f}_3}\Big/\frac{\bar{u}_1}{\bar{f}_1}}.$$

The "transport ratios" $\frac{\bar{u}_i}{\bar{f}_i}\Big/\frac{\bar{u}_1}{\bar{f}_1}$ are determined from the B.I.P.'s. The membranes investigated and the counter-ions used were the same again as in the preceding experiments (109).

For the bi-ionic systems the agreement is generally good, except when the concentration differences at both sides of the membrane are large.

For the tri- and quadri-ionic systems rather good agreement is also found in many cases. However, here again there are differences which are significantly larger than the reproducibility of the tests. The usability of formula (49) has also been investigated by F. Bergsma (12, 13) and A. J. Staverman (13). When formula (49) is correct, the relation between B.I.P. and activity of one of the solutions — at constant activity of the other solution — is graphically represented by a straight which intersects the x-axis with a slope of 58 mV per tenfold change of the activity. This is found to be correct in many cases. Exceptions are the cellophane type membranes; here the slope is often considerably less. In the investigation it was found that water transport is not the cause of the deviating behaviour. Increase of the liquid velocities to reduce the influence of the unstirred layers during the B.I.P. measurements did have some effect, but the deviation largely continued to exist, also at high stirring rates. However, it is possible that with strong hydrophylic membranes the influence of unstirred surface layers cannot be entirely suppressed. A second test of formula (49) is given by the following consideration:

When the correction terms of formula (48) can indeed be neglected, the "transport ratios" obtained from B.I.P. measurements must be the same as those from transport number measurements, the membrane being placed in a solution of both ions. This is found to be so as long as ions of the same kind are concerned, e.g. for Na^+ and K^+. For a larger difference between the ions, such as for the combinations Na^+—H^+, Na^+—Ag^+ and Ag^+—H^+ there are deviations.

From the results the conclusion was drawn that particularly in the interesting cases when specific selectivity occurs, the correction terms may be important and therefore formula (49) as such cannot be used for the derivation of transport ratios from B.I.P. measurements.

M. R. J. WYLLIE and S. L. KANAAN (185) also checked equation (49). From B.I.P. measurements they obtained the "transport ratios" $\frac{\bar{u}_1 \bar{f}_2}{\bar{u}_2 \bar{f}_1}$. For $\frac{\bar{f}_2}{\bar{f}_1}$ they substituted the constant of selectivity K_2^1, which follows from the Donnan relation

$$\frac{a_2}{a_1} = \frac{\bar{a}_2}{\bar{a}_1} = \frac{\bar{f}_2}{\bar{f}_1} \frac{\bar{c}_2}{\bar{c}_1} = K_2^1 \cdot \frac{\bar{c}_2}{\bar{c}_1} .$$

$\frac{\bar{u}_1}{\bar{u}_2}$ can then be calculated. These values were compared with those obtained experimentally with the aid of conductivity measurements. These investigators used heterogeneous membranes and various combinations of univalent counter-ions. In some cases the agreement was good. Rather large differences occurred too, maximum even of a factor of 3.

F. HELFFERICH and H. D. OCKER (54) have tested equation (51); they ignored the term indicating the change of the activity coefficient and took the constant of selectivity instead of $\frac{(\bar{f}_2^\omega)^{1/z_2}}{(\bar{f}_1^\omega)^{1/z_1}}$. The membrane investigated was of the homogeneous gel type with phenol sulfonic acid as a basis. For the combinations Na^+—H^+ and Na^+—K^+ a reasonably good agreement was found between the B.I.P.'s measured and the calculated ones. For the combination Na^+—Sr^{++} the difference was considerable. According to the authors, this was as could be expected, such in view of the simplifications applied.

4.2.3. Diffusion with Electric Current

The equations (64), (65), (66), (69) and (70) have been experimentally tested by P. MEARES and H. H. USSING (95) for a Zeo-Karb 315 membrane and NaCl solutions. $\bar{u}_{Na^+}$ was determined from self-diffusion measurements. At an external concentration up to 0.02 N the flux of the anions could be neglected and λ_{Na^+} was directly found from the conductivity of the membrane.

At 0.05 N the current transported by Cl^- ions was determined separately. The specific conductivity of the Na^+ ions in the membrane was determined from the conductivity of the membrane, taking into account the current transport of the Cl^- ions.

The value of ϱ_0 was determined from electro-osmosis measurements.

Equations (64) and (65) were found to satisfy only at 0.01 N solution. On the other hand there was reasonable agreement with equation (69) and (70) for all three concentrations measured, being 0.01 N, 0.02 N and 0.05 N. According to the authors the deviations from (64) and (65) are caused by the failure of equation (68).

The authors concluded that with the aid of the derived equations of the types (69) and (70) it is possible to differentiate between active and passive transport.

The results obtained might also serve as proof for the usability of the applied flux equations and of the membrane model which underlies the calculation. It remains remarkable, however, that equation (68) for the conductivity fails.

R. SCHLÖGL and U. SCHÖDEL (146) have supplied another proof for the usability of the Nernst-Planck flux equations combined with the M.S.T. model, also for the case that an electric current flows through an ion-selective membrane. They determined the concentration profiles of the mobile ions for the case of a cation selective membrane on the basis of phenol sulfonic acid and NaCl solutions, under application of an electric current.

In the Nernst-Planck equations used the activity coefficients were neglected; a term accounting for the electro-osmosis, however, is present. Calculated and measured concentration profiles could be made to inter-correspond by adapting the term for water transport. The values in-directly determined by electro-osmotic flow were now found to agree with those measured directly.

4.3. Conductivity

So far no theory has yet been developed which gives a complete description of the conductivity of a selective membrane, proceeding from the properties of the ions in free solution. The biggest difficulty is that of finding the relation between the apparent mobility of an ion in a membrane and the mobility in solution at infinite dilution. In § 5.1. chapter 3, a large number of factors have been given which are of influence. It is found in some cases that only the geometric factor and the tortuosity factor are of importance. D. MACKAY and P. MEARES (85) found that for the Zeo-Karb 315 membrane in NaCl solutions the apparent mobility of the counter-ions at high external concentration is determined by the two factors referred to above, calculated according to equations (73) and (78). This was also the case for the co-ions, independent of the external concen-tration. It must be pointed out, however, that the membrane investigated possesses a high water content and a low capacity. Thus the influence of the two factors is relatively small. Proceeding from the mobilities of the ions in the membrane, which can be determined from measurements of coefficients of self-diffusion, it is possible to test experimentally equations for the conductivity of the membrane, such as equation (72). This has inter alia been done by D. MACKAY and P. MEARES (*85*). They checked the validity of the formula:

$$\bar{\varkappa} = F\left\{\frac{F}{RT}\left(\bar{D}_{Na^+}\bar{C}_{Na^+} + \bar{D}_{Cl^-}\bar{C}_{Cl^-}\right) + \nu\left(\bar{C}_{Na^+} - \bar{C}_{Cl^-}\right)\right\}. \tag{91}$$

Equation (91) is obtained from (72) by substituting the Fokker-Einstein relation $\bar{u}_i = \frac{F}{RT}\bar{D}_i$ and the expression for the osmotic flow:

$$\nu = -\frac{F\bar{A}'}{w\varrho_0}\frac{\Delta\varphi}{d}.$$

In the case of a Zeo-Karb 315 membrane in equilibrium with NaCl solutions, equation (91) was found to be satisfied at an external concentration of 0.01 N and also at 1 N. Between the two concentrations the calculated values are higher than the measured values. This was ascribed by D. MACKAY and P. MEARES to a non-uniform distribution of charge.

At low salt concentrations, the share of the co-ions in the conductivity is negligible and thus equation (68) is obtained. As was said before, it was found by P. MEARES and H. H. USSING (*95*) that, strangely enough, this formula does not hold true.

In general it must be remarked that more experimental material is essential. For the case that the permeability to water is low, the following approximation applies:

$$\bar{\varkappa} = \Sigma_i z_i^2 F \bar{c}_i \bar{u}_i. \tag{92}$$

This equation has inter alia been tested by G. MANECKE and K. F. BON-HOEFFER (*89*) and by G. MANECKE (*90*).

The authors first mentioned investigated an anion-exchange membrane consisting of polyethylene imine crosslinked with epichlorohydrin in equilibrium with KCl solutions. The concentration of Cl^- ions and K^+ ions in the membrane were determined analytically. The mobility of the Cl^- ions in the membrane was determined according to the principle of the moving boundary. Thus in a membrane strip a sharp boundary is formed between the OH^- and Cl^- ions. This boundary will move under the influence of an electric field. This may be made visible with the aid of an indicator. The speed of the boundary is determined by the slowest ion, by consequence the Cl^- ion in this instance.

The determination with the aid of radioactive isotopes has been mentioned too. It was assumed that the ratio of the mobilities of K^+ ions and Cl^- ions is the same as in aqueous solution. Satisfactory agreement was found to exist between the calculated and the measured conductivity.

Similar results were obtained by G. MANECKE (*90*) with cation exchange membranes on a phenol sulfonic acid basis and styrene sulfonic acid basis.

5. Applications of Ion-Exchange Membranes

To enable an impression of the large number of possible applications of ion-exchange membranes, a brief survey is here given of the pertinent literature. The major application, which has already been implemented on a technical scale, is the electrodialytical desalting of brackish water in order to obtain drinkingwater (19, 64, 65, 66, 177, 178, 179, 180).

Deionization of different solutions can be performed with a high current efficiency (15, 25).

A peculiar method of desalting water is the osmionic demineralization developed by G. W. Murphy (102, 103). In a multicell are circulated side by side a very concentrated salt solution, a moderately concentrated salt solution and a dilute salt solution to be desalted. The arrangement is now such that only Na^+ and Cl^- ions can move from the brine to the moderately concentrated salt solution, when simultaneously also Na^+ and Cl^- ions can move from the dilute salt solution to this moderately concentrated solution. In this manner the dilute solution is desalted.

The recovery of important metals or their salts is possible by electrolysis in cells provided with ion-selective membranes, e.g. of uranium (71, 72, 73, 75), of magnesium from sea water (130), of iodine from iodide containing brines (158), of manganese (74).

A number of separations can be performed with the aid of ion-exchange membranes. The separation of electrolytes and non-electrolytes was investigated by G. Manecke and H. Heller (92).

An electro-osmotic ion separation method was patented by D. R. Dewey and E. R. Gilliland (37).

It is based on a countercurrent-flow ion-migration through a series of membranes. A new electrolytic separation technique, based on the ion-exclusion principle is described by E. Glueckauf and G. P. Kitt. They used an acid-base double membrane to separate uni-, di- and trivalent ions. This double membrane is so placed that the anion exchange side faces the anode and the cation exchange side faces the cathode. The salt ions can now only pass through the double membrane, if they first move part of the way as co-ions.

Separations are now effected, because the concentrations of the various co-ions in the membrane are generally different; e.g. for a univalent co-ion this concentration is higher than for a bivalent co-ion. The current efficiency is low due to the large contribution of H^+ and OH^- ions to the current transport, especially at higher current densities.

K. Bril, S. Bril and P. Krumholz (24) developed a method to separate rare earth mixtures with the help of ion-exchange membranes. The method is based on the difference in stability of complexes of rare earths. When a mixture of rare earths with a complex former (e.g. ethylenediamine tetra acetic acid) is placed in an electro-dialysis cell, the metal ions are removed through the negative membrane and the negatively charged complexes disappear through the positive membrane. If the stability constants are different, separation of the rare earths occurs.

D. LOGIE (83) described a new analytical separation technique by applying ion-exchange membranes, which can be used for the determination of boron in sodium metal. By treatment with water, the Na is converted to NaOH, borate being formed from the boron. When the solution is introduced in the anode chamber of a two-cell apparatus fitted with a negative membrane, the Na^+ ion is transported to the cathode chamber, whereas the borate anion remains in the anode chamber. In general this method can be applied, if the trace element yields an ion with a charge which opposite to that of the main component.

Focusing ion-exchange is applied by E. SCHUMACHER et al. (151—154) to separate different ions inter alia: Co, Cu, Ba, Sr and rare earth ions; Cu^{++}, UO_2^{++} and Sr^{++}.

The apparatus consists of a system of channels separated by selective membranes. If a direct current is maintained across a stationary gradient of complex-forming anions, a focusing of the position of the cations is produced.

The arrangement is such that at the cathode side the highest concentration of complex-forming anions is present. Here the resulting complex has a negative charge and will move in the direction of the anode. At the anode side the concentration of complex-forming anions is lowest. Here the complex-ion will have a positive charge and move towards the cathode. The ions of the element concerned will collect at the place where the charge of its complex just equals zero.

Also a p_H gradient can be used in some cases. The purpose of using selective membranes is mainly that of avoiding convection, so that the gradients are not disturbed.

In a multicell provided with ion-selective membranes one can perform double conversions according to the scheme (14, 67): $M^+S^- + N^+Z^- \rightarrow M^+Z^- + N^+S^-$ (see Fig. 2).

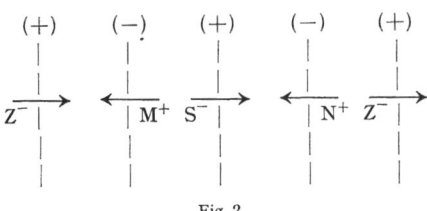

Fig. 2

Examples are the preparation of sodium hydroxide from sodium-chloride and lime-milk; the preparation of sodiumhydroxide and ammoniumchloride from sodiumchloride and ammoniumhydroxide (59); the conversion of uranylnitrate to uranic fluoride salts (150); the electrolytic production of weak acids (131), according to the scheme

$$NaZ + HCl \rightarrow HZ + NaCl ;$$

and the electrolytic production of sodiumhydroxide, sodiumcarbonate and sodiumbicarbonate from seawater and lime-milk (106).

Acids and bases can be obtained if the electrode reactions are involved; for example the electrolytic conversion of ammoniumsulfate in ammoniumhydroxide and sulfuric acid, using a positive membrane has been patented (129).

Another example is the production of sodiumhydroxide, starting from sodiumchloride (78).

Also is claimed the electrolytic conversion of sodiumchloride in sodiumhydroxide and chlorine (18). In all these instances, the selective membrane is applied in order to increase the current efficiency by either impeding the disappearance of OH^- ions from the cathode cell to the anode cell, or that of H^+ ions in the reverse direction.

Another way to obtain acids and bases is to make use of the transport of hydrogen ions through cation-exchange membranes, respectively of hydroxyl ions through anion-exchange membranes, which occurs during electrodialysis if no new counterions are offered by the depleted solution (1, 113, 114, 115).

Electrodialysis with ion-exchange membranes can also be used to concentrate electrolytes. In Japan this method meets with much interest for the recovery of sodiumchloride from seawater (104, 105, 156, 172, 188).

Ion-selective membranes can also be used as electrode-membranes to determine the activity of ions in solution (2, 8, 9, 27, 122, 176).

Here it is assumed that the membrane is 100% selective, so that the membrane potential is as follows:

$$E_M = \frac{RT}{F} \ln \frac{a_2}{a_1} . \tag{46}$$

The membrane is placed between a solution of known activity and a solution whose activity is to be measured. The latter activity is found by measuring the membrane potential.

Ch. W. Carr investigated in this manner the binding of small ions in protein solutions (26).

The method is also used to measure the activity in precipitates and suspensions (140). The possibilities of application of the method are limited, because the membranes have little specific selectivity. Hence the membrane potential is determined by the activity of all the counter-ions together and not by the activity of one special ion, e.g. H^+ ions in the instance of the glass electrode.

St. W. Feldberg and Cl. E. Bricker describe the use of ion-exchange membranes in coulometry (39).

Ion-exchange membranes can be applied in primary cells (47), in rechargeable batteries (82) and in low-temperature fuel cells (48).

Ion-exchange membranes also show some promise in the solution of waste problems, inter alia the treatment of spent pickle liquors by electrodialysis is discussed (80). A very high degree of deionization can be achieved with ion-selective membranes. P. Cohen investigated the electrodialysis of simulated pressurized water reactor coolant (34). The specific resistance of the treated water was as low as 0.5 to 3,0 MΩ cm.

Appendix 1. List of Symbols

χ_i = generalised force
Φ_i = flux of species i
L_{ik} = phenomenological coefficient
E = electrical potential difference
ΔP = pressure difference
L_E = electrical conductivity of one square cm of membrane
n = number of components of a system
$\tilde{u}_i$ = velocity of component i
Ω_{ij} = friction coefficients
T_i = electrical transport number
t_i = reduced transport number or transference number
z_i = charge of ion i
v_i = partial volume of ion i per gmol
v_0 = partial volume of the solvent per gmol
a_i = activity of ion i
c_i = concentration of ion i
f_i = activity coefficient of ion i
μ_i = chemical potential of ion i per gmol
u_i = mobility of ion i
D_i = diffusion coefficient of ion i

$\bar{A}$ = concentration of fixed charges per ml solution
$\bar{A}'$ = concentration of fixed charges per unit volume of the swollen membrane
d = thickness of the membrane
δ = thickness of the unstirred film
ω = sign of charge of the fixed groups
w = relative volume of the pores of the membrane
ϱ_0 = flow resistance of the membrane
$\bar{u}_w$ = mobility of the water in the membrane
$\bar{\varkappa}$ = specific electrical conductivity of the membrane
D_H = hydrodynamic water permeability
ν = velocity of the centre of gravity
φ = electrical potential
F = Faraday's constant
E_s = streaming potential
i_s = streaming current.

M.S.T.-model = Meyer-Sievers-Teorell-model (see Introduction, sec. 1).

Literature

1. Arnold, H. W., and G. P. Monet: (to E. I. du Pont de Nemours & Co.) U.S 2,721,171. Octobre 18 (1955).
2. Asahi Glass Co.: Brit. 793, 212, April 9 (1958).
3. Asahi Glass Co.: Brit. 804, 176.
4. Ashida, K. (Hodogaya Chem. Co.): Japan 8608 (1957).
5. Aten, A. H. W.: Chem. Weekblad 25, 211 (1928).
6. Atsugi, T., M. Ichikawa and M. Yamada (to Tokuyama Soda Co.): Japan 4596 (1957), July 6.
7. — — — (to Tokuyama Soda Co.): Japan 2293 (1958).
8. Basu, A. S.: Science and Culture (India) 21, 447 (1956).
9. — J. Indian Chem. Soc. 35, 451 (1958).
10. Farbenfabriken Bayer Akt.Ges., Brit. 783, 450.

11. Farbenfabriken Bayer Akt.-Ges., Ger. 958, 458.
12. Bergsma, F.: A study of bi-ionic membrane potentials. Thesis, Leiden (1957).
13. — and A. J. Staverman: Disc. Faraday Soc. 21, 61 (1956).
14. — (To the Nederlandse Organisatie voor toegepast natuurwetenschappelijk onderzoek T.N.O.) Dutch 86, 370, Octobre 15 (1957); Brit. 795, 636, May 28 (1958).
15. Block, R. J., and W. H. Wingerd: U. S. 2,830,905, April 15 (1958).
16. Bochove, C. v., and H. G. Roebersen: (to the Nederlandse Organisatie voor toegepast natuurwetenschappelijk onderzoek T.N.O.), Ger. 1,000,350, Jan. 10 (1957).
17. Bodamer, G. W.: (to Rohm & Haas Co.): U. S. 2,681,320, June 15 (1954).
18. — (to Rohm & Haas Co.): U. S. 2,827,426, March 18 (1958).
19. Boer-Nieveld, Y., and D. Pauli: Survey of water desalting investigations, in particular the electrodialytic method. Rapport T. A. 270 (Alg. Techn. Afd. T.N.O., Den Haag, Oct. 1952).
20. Bonhoeffer, K. F., L. Miller and U. Schindewolf: Z. physik. Chem. 198, 270 (1951).
21. — and U. Schindewolf: Z. physik. Chem. 198, 281 (1951).
22. Boyd, G. E., B. A. Soldano and O. D. Bonner: J. Phys. Chem. 58, 456 (1954).
23. Bradfield, R.: J. Phys. Chem. 33, 1724 (1929).
24. Bril, K., S. Bril and P. Krumholz: J. Phys. Chem. 63, 256 (1959).
25. Buriánek, J. and D. Šlechtová: Listy cukrovar 75, 62, 66 (1959).
26. Carr, Ch. W.: Arch. Biochem. Biophys. 62, 476 (1956).
27. Chaussidon, J.: Compt. rend. 244, 2798 (1957).
28. Chen, W. K. H., R. B. Mesrobian, D. S. Ballentine, D. J. Metz and A. Glines: J. Polymer Sci. 23, 903 (1957).
29. Clarke, J. T. (to Ionics Inc.): U. S. 2,731,408, Jan. 17 (1956).
30. — (to Ionics Inc.): U. S. 2,731,411.
31. — (to Ionics Inc.): U. S. 2,732,350, Jan. 24 (1956).
32. — (to Ionics Inc.): U. S. 2,756,202, July 24 (1956).
33. — (to Ionics Inc.): U. S. 2,800,445, July 23 (1957).
34. Cohen, P. (Westinghouse Bettis Plant): Ind. Eng. Chem. 51, 66 (1959).
35. Cosgrove, J. D., and J. D. H. Strickland: J. Chem. Soc. (London) 1950, 1845
36. Danielli, J. F.: in: The permeability of natural membranes, by Davson and Danielli (Cambridge 1943), chap. XXI and appendix A.
37. Dewey, D. R., and E. R. Gilliland (to Ionics Inc.): U. S. 2,741,591.
38. Dray, S., and K. Sollner: Biochem. et Biophys. Acta 22, 213, 220 (1956).
39. Feldberg, St. W., and Cl. E. Bricker: Anal. Chem. 31, 1852 (1959).
40. Glasstone, S., K. J. Laidler and H. Eyring: The theory of rate processes, chapter X. New York: Mc.Graw-Hill Book Co. 1941.
41. Graydon, W. F., and R. J. Stewart: J. Phys. Chem. 59, 86 (1955).
42. — U. S. 2,877,191, March 3 (1959).
43. Gregor, H. P., and D. M. Wetstone: Disc. Faraday Soc. 21, 162 (1956).
44. — H. Jacobson, R. C. Shair and D. M. Wetstone: J. Phys. Chem. 61, 141 147, 151 (1957).
45. — and D. M. Wetstone: Z. Elektrochem. 62, 274 (1958).
46. Groot, S. R. de: Thermodynamics of irreversible processes. Amsterdam: North-Holland Publishing Company 1951.
47. Grubb, W. T. (to Gen. Electric Co.): U. S. 2,861,116, Novembre 18 (1958).
48. — and L. W. Niedrach: J. Electrochem. Soc. 106, 275 (1959); J. Electrochem. Soc. 107, 131 (1960).
49. Guggenheim, E. A.: J. Phys. Chem. 34, 1758 (1930).

50. HAZENBERG, J. F. A., and B. P. KNOL (to T.N.O.): South Africa Application no. 1754/58.
51. HELFFERICH, F.: Z. physik. Chem. N. F. **4**, 386 (1955).
52. — Disc. Faraday Soc. **21**, 83 (1956).
53. — and R. SCHLÖGL: Discussionsremark. Disc. Faraday Soc. **21**, 133 (1956).
54. — and H. D. OCKER: Z. physik. Chem. N. F. **10**, 213 (1957).
55. — Ionenaustauscher, Vol. 1, chapt. 8. Weinheim: Verlag Chemie 1959.
56. HONEY, E. M. O., and CH. R. HARDY (to Chloride Electrical Storage Co. Ltd.): U. S. 2,810,932, Oct. 29 (1957).
57. I. C. I. Belgium 565,980, April 15 (1958).
58. I. C. I. Australia Application, no. 43,088/58.
59. ISHIBASHI, N., and N. EMURA: Denki Kagaku **25**, 625 (1957).
60. JONG, G. J. DE (to Stamicarbon N. V.): U. S. 2,858, 264.
61. JUDA, W., and W. A. McRAE (to Ionics Inc.): U. S. 2,636,851, April 28 (1953).
62. — and A. A. KASPER (to Ionics Inc.): U. S. 2,731,425, Jan. 17 (1956).
63. KAHLWEIT, M.: Z. Physik. Chem. N. F. **6**, 45 (1956).
64. KATZ, WM. E.: Proc. Ann. Conf. Maryland-Delaware. Water and Sewage Assoc. **29**, 36 (1956).
65. — Chem. Eng. Progr. **53**, 140 (1957).
66. KIRKHAM, T. A.: Chem. Eng. **63**, 185 (1957).
67. KOLLSMAN, P.: U. S. 2,835,632, May 20, (1958); U. S. 2,835,633.
68. KRESSMAN, T. R. E.: Nature (London) **165**, 568 (1950).
69. — J. Appl. Chem. **4**, 123 (1954).
70. — and S. C. SMITH (to Permutit Co. Ltd.): B. P. 810,391, March 18, (1959).
71. KUNIN, R.: U. S. 2,733, 200, Jan. 31 (1956).
72. — U. S. 2,739,934, March 27 (1956).
73. — U. S. 2,741,589, April 10 (1956).
74. — U. S. 2,741,590, April 10, (1956).
75. — U. S. 2,832,727, April 29 (1958).
76. KUWATA, Ts., and S. YOSHIKAWA (to Asahi Glass Co.): Japan 4590 (1957), Juli 6.
77. — — (to Asahi Glass Co.): Japan 4591 (1957).
78. — — H. MARUYAMA, SH. UEMURA and T. YAWATAYA (to Asahi Glass. Co.): Japan 9560 Nov. 14 (1957).
79. — — S. SEKINO, E. NISHIHARA, and Y. MINEKI (to Asahi Glass Co.): Japan 7840, Sept. 5 (1958).
80. LEWIS, D. J., and F. L. TYE: J. Appl. Chem. **9**, 279 (1959).
81. LEWIS, M., and K. SOLLNER: J. Electrochem. Soc. **106**, 347 (1959).
82. LINDSAY, P.: Howard I. R. E. Natl. Conv Record 4, Pt. 6,132 (1956).
83. LOGIE, D.: Chem. and Ind. (London) **1957**, 225.
84. LORIMER, J. W., E. I. BOTERENBROOD and J. J. HERMANS: Disc. Faraday Soc. **21**, 141 (1956).
85. MACKAY, D., and P. MEARES: Trans Faraday Soc. **55**, 1221 (1959).
86. MACKIE, J. S., and P. MEARES: Proc. Roy. Soc. (London), A. **232**, 485 (1955).
87. — — Proc. Roy. Soc. (London) A. **232**, 498 (1955).
88. — — Proc. Roy. Soc. (London) A. **232**, 510 (1955).
89. MANECKE, G., and K. F. BONHOEFFER: Z. Elektrochem. **55**, 475 (1951).
90. — Z. physik. Chem. **201**, 193 (1952).
91. — (Farbenfabriken Bayer A. G.): Ger. 1,020,600, December 12 (1957).
92. — and H. HELLER: Z. Elektrochem. **61**, 150 (1957).
93. MARSHALL, C. E.: J. Phys. Chem. **48**, 67 (1944).

94. Meares, P., and H. H. Ussing: Trans. Faraday Soc. **55**, 142 (1959).
95. — — Trans. Faraday Soc. **55**, 244 (1959).
96. — Trans. Faraday Soc. **55**, 1970 (1959).
97. Meyer, K. H., and J. F. Sievers: Helv. Chim. Acta **19**, 649, 665, 987 (1936); Helv. Chim. Acta **20**, 634 (1937).
98. — — Helv. Chim. Acta **19**, 665 (1936).
99. — and W. Strauss: Helv. Chim. Acta **23**, 795 (1940).
100. — and P. Bernfeld: Helv. Chim. Acta **28**, 962 (1945).
101. Motozato, Y., H. Egawa, H. Maegaki and K. Kunitake: Kôgyô Kagaku Zasshi **59**, 479 (1956).
102. Murphy, G. W.: Ind. Eng. Chem. **49**, 28 A (1957).
103. — Ind. Eng. Chem. **50**, 1181 (1958).
104. Nakazawa, H. (to Tokuyama Soda Co.): Japan 2615, April 21 (1955); Japan 5415, July 24 (1957).
105. — T. Atsugi and K. Onoe (to Tokuyama Soda Co.): Japan 4026, June 14 (1955).
106. — and K. Ogami (to Tokuyama Soda Co.): Japan 3960, June 20 (1957).
107. — K. Onogami and Y. Mizutani (to Tokuyama Soda Co.): Japan 91, Jan. 17 (1959).
108. Neihof, R.: J. Phys. Chem. **58**, 916 (1954).
109. — and K. Sollner: Disc. Faraday Soc. **21**, 94 (1956).
110. Nernst, W.: Z. physik. Chem. **2**, 613 (1888); **4**, 129 (1889).
111. Nishihara, A., Y. Mineki and M. Sekino (to Asahi Glass Co.): Repts. Research Lab. Asahi Glass. Co. **6**, 20 (1956).
112. Nishimura, M., et al. (to Osaka Municipal Ind. Research Inst.): Kagaku to Kôgyô (Osaka) **32**, 354 (1958).
113. Oda, K., and M. Murakoshi (to Noguchi Research Institute, Inc.): Japan 3962, June 20 (1957).
114. — — (to Noguchi Research Institute Inc.): Japan 2023 (1958); March 25; Japan 2469, April 10 (1958).
115. — — (to Noguchi Research Institute, Inc.): U. S. 2,829,095, April 1 (1958).
116. — Sh. Matsuda and T. Saito (to Noguchi Research Institute, Inc.): Japan 9489, Oct. 25 (1958); Japan 9490, Oct. 25 (1958); Japan 9491, Oct. 25 (1958).
117. Oel, H. J.: Z. physik. Chem. N. F. **5**, 32 (1955).
118. — Z. physik. Chem. N. F. **15**, 280 (1958).
119. Ogami, K., and C. Kanda (to Tokuyama Soda Co.): Japan 1897, March 19 (1958).
120. Ongaro, D. (to Montecatini): Ital. 529,967, June 30 (1955).
121. Overbeek, J. Th. G.: J. Colloid Sci. **8**, 593 (1953).
122. Patnode, H. W., and M. R. J. Wyllie (to Gulf Research & Development Co.): U. S. 2,614,976, Oct. 21 (1952).
123. Partridge, S. M., and A. M. Peers: J. Appl. Chem. **8**, 49 (1958).
124. Peers, A. M.: Disc. Faraday Soc. **21**, 124 (1956).
125. Permutit Co. Ltd.: Brit. 747, 948, July 12 (1954).
126. Peterson, M. A., and H. P. Gregor: J. Electrochem. Soc. **106**, 1051 (1959).
127. Planck, M.: Ann. Phys. u. Chem. **39**, 161 (1890).
128. Pritchett and Gold Ltd.: B. P. 743,926, Jan. 1 (1956).
129. Rohm & Haas Co.: B. P. 738,519, Oct. 12 (1955).
130. Rohm & Haas Co.: B. P. 738,520, Oct. 12 (1955).
131. Rohm & Haas Co.: B. P. 767,103, Jan. 30 (1957).

132. Rohm & Haas Co.: B. P. 782,059, Aug. 28 (1957).

133. ROMANKEVICH, M. YA.: Ukrain. Khim. Zhur. **24**, 325 (1958).

134. ROSS, S. D., and W. W. SCHROEDER JR.: U. S. 2,648,717, Aug. 11 (1953).

135. South Africa Council for Scientific and Industrial Research (S.A.C.S.I.R.), Fr. 1,163,010, Sept. 22 (1958).

136. South Africa Council for Scientific and Industrial Research (S.A.C.S.I.R.), Brit. 813,601, May 21 (1959).

137. SCATCHARD, G.: J. Am. Chem. Soc. **75**, 2883 (1953).

138. — Ion transport across membranes. New York: H. T. Clarke Editor, Academic press. Inc. 1954.

139. — and F. HELFFERICH: Disc. Faraday Soc. **21**, 70 (1956).

140. SCHEFFER, F., u. F. LUDWIG: Atompraxis **4**, 331 (1958).

141. SCHINDEWOLF, U., u. K. F. BONHOEFFER: Z. Elektrochem. **57**, 216 (1953).

142. SCHLÖGL, R., u. F. HELFFERICH: Z. Elektrochem. **56**, 644 (1952).

143. — Z. Elektrochem. **57**, 195 (1953).

144. — Z. physik. Chem. N. F. **1**, 305 (1954).

145. — Z. physik. Chem. N. F. **3**, 73 (1955).

146. — and M. SCHÖDEL: Z. physik. Chem. N. F. **5**, 372 (1955).

147. — and B. STEIN: Z. physik. Chem. N. F. **13**, 111 (1957).

148. SCHMID, G.: Z. Elektrochem. **56**, 181 (1952).

149. SCHORS, A., J. H. V. D. NEUT, H. G. ROEBERSEN and O. P. V. D. WERFF (to T.N.O.): Belgium 537,438; Brit. 778,001.

150. SCHULZ, W. W., E. W. NEUVAR, J. L. CARROLL and R. E. BURNS (Gen. Elec. Co., Wash.): Ind. and Eng. Chem. **50**, 1768 (1958).

151. SCHUMACHER, E.: Helv. Chim. Acta **40**, 221, 228, 234 (1957).

152. — Helv. Chim. Acta **40**, 2322 (1957).

153. — and H. J. STREIFF: Helv. Chim. Acta **41**, 824 (1958).

154. — and R. FLÜHLER: Helv. Chim. Acta **41**, 1572 (1958).

155. SEKINO, SH., A. NISHIHARA and Y. MINEKI (to Asahi Glass Co.): Japan 1145, Febr. 21 (1958).

156. — H. HANI, M. YAMADA and SH. WADA (to Asahi Glass Co.): Japan 2468, April 10 (1958).

157. — et al. (to Asahi Glass Co.): Japan 645 (1959).

158. SHIMIZU, H., Y. ARAI, A. SATO and Y. DOCHI (to Nippon Organo Co.): Japan 58, Jan. 17 (1959).

159. SKOGSEID, A.: Thesis. Oslo 1948.

     — U. S. 2,592,350 (1952).

160. SLOUGH, W.: Trans. Faraday Soc. **55**, 1030, 1036 (1959).

161. SOLDANO, B. A., and G. E. BOYD: J. Am. Chem. Soc. **75**, 6107 (1953).

162. SOLLNER, K.: J. Electrochem. Soc. **97**, 139c (1950).

163. — S. DRAY, E. GRIM and R. NEIHOF: in Ion transport across membranes. page 144. New York: H. T. Clarke, Editor. Academic Press Inc. 1954.

164. SPIEGLER, K. S.: Trans. Faraday Soc. **54**, 1408 (1958).

165. STAVERMAN, A. J.: Chem. Weekblad **47**, 1 (1951).

166. — Trans. Faraday Soc. **48**, 176 (1952).

167. STEWART, R. J., and W. F. GRAYDON: J. Phys. Chem. **61**, 164 (1957).

168. TEORELL, T.: Proc. Soc. Expt. Biol. Med. **33**, 282 (1935).

169. TEYSSIÉ, PH., and H. P. GREGOR: Makromol. Chem. **31**, 192 (1959).

170. TITOV, V. S., U.S.S.R. 115,837, Nov. 22 (1958).

171. — U.S.S.R. 115,677, Nov. 29 (1958).

172. Tsunoda, Y., and M. Seko (to Asahi Chem. Ind. Co.): Japan 5068, July 23 (1955); Japan 5069, July 23 (1955).
173. — N. Seko, M. Watanabe, A. Ehara and T. Misumi (to Asahi Chem. Ind. Co.): Japan 4142, June 24 (1957).
174. — — — — — (to Asahi Chem. Ind. Co.): Japan 4143, June 24 (1957); Japan 4145, June 24 (1957).
175. — — et al. (to Asahi Chem. Ind. Co.): U. S. 2,894,917.
176. — — (to Asahi Chem. Ind. Co.): U. S. 2,864,776, Dec. 16 (1958).
177. Volckman, O. B.: Brit. Chem. Eng. 2, 146 (1957).
178. Wegelin, E.: Bull C.B.E.D.E. nr. 21-1953/III, 182.
179. Whalley, C. H. de: Chem. and Ind. Jan. 4, 8 (1958).
180. Wiechers, S. G., and C. van Hoek: Research 6, 194 (1953).
181. Winger, A. G., G. W. Bodamer and R. Kunin: J. Electrochem. Soc. 100, 178 (1953).
182. Woermann, D., K. F. Bonhoeffer u. F. Helfferich: Z. physik. Chem. N. F. 8, 265 (1956).
183. V. E. B. Farbenfabrik Wolfen, Ger. (East) 13,622. Suppl. to Ger. (East) 13,389, June 25 (1957).
184. Wyllie, M. R. J., and H. W. Patnode: J. Phys. & Colloid Chem. 54, 204 (1950).
185. — and S. L. Kanaan: J. Phys. Chem. 58, 67 (1954).
186. — U. S. 2,774,108, Dec. 18 (1956).
187. — (to Gulf Research & Develp. Co.): U. S. 2,820,756, Jan. 21 (1958).
188. Yawataya, T., and Sh. Uemura (to Asahi Glass Co.): Japan 3213, May 14 (1955).
189. — and H. Ukihashi (to Asahi Glass Co.): Japan 4111, June 24 (1957).
190. Zwolinski, B. J., H. Eyring and C. E. Reese: J. Phys. Chem. 53, 1426 (1949).

Fortschr. Hochpolym.-Forsch., Bd. 2, S. 363—400 (1961)

Anwendung und Ergebnisse
der Röntgenkleinwinkelstreuung in festen Hochpolymeren

Von

G. POROD

Institut für physikalische Chemie der Universität Graz

Mit 8 Abbildungen

Inhaltsverzeichnis

Seite

I. Theorie und Methode 363
   1. Allgemeine Prinzipien der Kleinwinkelstreuung 363
   2. Verdünnte Systeme 367
     a) Auswertung nach Größe und Form 367
     b) Auswertung nach Polydispersität 370
   3. Dichtgepackte Systeme 372
     a) Unorientierte Systeme, diffuse Kleinwinkelstreuung 373
     b) Orientierte Systeme, diffuse Streuung 376
     c) Geordnete Systeme, diskrete Streuung 378
II. Ergebnisse an festen Hochpolymeren 380
   1. Unorientierte Präparate 380
   2. Orientierte Systeme, meridiale Kleinwinkelstreuung 382
     a) Meridianreflexe 382
     b) Schichtkristalle 385
     c) Deutung der Meridianreflexe 387
     d) Zusammenhang der Meridianreflexe mit dem Weitwinkeldiagramm. 390
   3. Orientierte Systeme, äquatoriale Kleinwinkelstreuung 392
     a) Äquatoriale Streukurven 392
     b) Hohlräume; Auswertung der Streukraft 394
Zusammenfassung 396
Literatur 397

I. Theorie und Methode

1. Allgemeine Prinzipien der Kleinwinkelstreuung

Wie das sichtbare Licht werden auch Röntgenstrahlen an Inhomogenitäten der Materie abgebeugt. Wenn Inhomogenitäten in kolloiden Dimensionen vorliegen, die ein Vielfaches der praktisch verwendeten Röntgenwellenlängen betragen, erstreckt sich der Effekt nur zu kleinen Winkeln — bis zu wenigen Winkelgraden — und wird daher als

Röntgenkleinwinkelstreuung oder kurz als Kleinwinkelstreuung (KWS) bezeichnet. Sie hat in letzter Zeit als eine wertvolle Methode zur Erforschung kolloider Systeme eine immer steigende Bedeutung gewonnen.

Die Beugung eines einfallenden Röntgenstrahlenbündels, des sog. Primärstrahls, erfolgt fast ausschließlich an den Elektronen. Wir können uns vorstellen, daß diese kohärente Sekundärwellen aussenden, die sich in einem Beobachtungspunkt wieder zu einer resultierenden Amplitude überlagern. Dabei treten Phasendifferenzen auf, die mit zunehmendem Streuwinkel zu einer immer stärkeren Schwächung der abgebeugten Intensität und schließlich zu ihrer fast völligen Auslöschung führen. Die nähere Art und Weise dieses Interferenzvorganges hängt von Größe, Form und Anordnung der kolloiden Teilchen, also der gesamten morphologischen Struktur des Systems ab. Das Problem besteht nun darin, aus der Kleinwinkelstreuung auf die Struktur rückzuschließen. Wir haben dabei wohl zu unterscheiden zwischen der sog. Absolutintensität der KWS und ihrer Winkelabhängigkeit, die als Streukurve bezeichnet wird. Diese beiden Daten weisen auf völlig verschiedene Eigenschaften des Systems hin und können unabhängig voneinander diskutiert werden. Doch gibt es auch Größen, die erst aus der Kombination beider erschlossen werden können.

Die KWS ist unabhängig von der atomaren Struktur der Materie. Es besteht daher in dieser Hinsicht kein Unterschied zwischen kristallinen und amorphen Teilchen. Man kann sich innerhalb einer homogenen Phase die Elektronen verschmiert denken und durch die Angabe einer Elektronendichte (üblicherweise in Mol/cm³) beschreiben. Da ein völlig homogenes Präparat keine KWS gibt, kommt es für den Effekt nur auf die Abweichungen der Elektronendichte vom Mittelwert, bzw. bei einem Zweiphasensystem auf die Differenz an. Es gilt dabei das Reziprozitätsgesetz der Optik (Babinetsches Theorem), d. h. die KWS ändert sich nicht, wenn etwa in einem Zweiphasensystem die Elektronendichten vertauscht werden. Aus der KWS allein geht daher nicht hervor, ob es sich um Teilchen oder um Löcher handelt. Doch wird diese Frage in den meisten Fällen aus der Natur des untersuchten Systems beantwortet werden können. In Zweifelsfällen ist eine Unterscheidung durch eine geeignete Variation des Präparates (Quellen, Änderung der Elektronendichte des Mediums u. dgl.) möglich.

Auch abgesehen von der oben erwähnten Unbestimmtheit ist die KWS nicht eindeutig auswertbar. Man wird es plausibel finden, daß die unendliche Mannigfaltigkeit der kolloiden Systeme durch die Schar der meist nicht sehr typischen Streukurven nicht eindeutig wiedergegeben wird. Systeme, die eine identische Streukurve liefern, wollen wir als „streuungsäquivalent" bezeichnen. Viel unangenehmer ist aber noch die praktische Mehrdeutigkeit, die auf dem Umstand beruht, daß die KWS experimentell nicht im ganzen Winkelbereich und nur mit einer Genauigkeit von einigen Prozent bestimmt werden kann. In der Praxis müssen daher auch solche Systeme als streuungsäquivalent aufgefaßt werden, zwischen denen innerhalb der experimentellen Genauigkeit einer Kleinwinkelaufnahme nicht unterschieden werden kann. Doch besteht auch hier die Möglichkeit, diese Mehrdeutigkeit durch eine passende Veränderung des Systems wieder aufzuheben. Die dazu nötigen Maßnahmen werden natürlich von Fall zu Fall verschieden sein.

Die Winkelausdehnung der KWS ist grundsätzlich wie bei allen optischen Beugungserscheinungen der Größe der Inhomogenitäten antibat. Eine lineare Vergrößerung eines kolloiden Systems um einen bestimmten Faktor würde bewirken, daß die Streukurve um den gleichen Faktor zu kleineren Winkeln gestaucht wird. Darauf beruht die Möglichkeit, aus der KWS zu Aussagen über eine Teilchengröße zu gelangen. Maßgebend ist dabei stets die Abmessung des Teilchens in der Abbeugungsrichtung. Wenn daher, wie dies gerade bei den festen Hochpolymeren in der Regel der Fall ist, anisometrische Inhomogenitäten vorliegen, die nach einer Richtung orientiert sind, entsteht auch ein anisometrisches Kleinwinkeldiagramm (oval, strich- oder besenförmig), das roh die Teilchenform wiederspiegelt, allerdings um 90° verdreht, denn der kleinen Dimension entspricht ja die große Ausdehnung der KWS und umgekehrt.

Eine solche Anisometrie läßt sich allerdings nur dann beobachten, wenn man für die Aufnahme eine Lochkamera verwendet, etwa in der von KIESSIG (1, 2) angegebenen Konstruktion. In der Regel benützt man aber für Kleinwinkeluntersuchungen einen strichförmig ausgeblendeten Primärstrahl. Denn nur mit einem solchen läßt sich gleichzeitig eine hohe Auflösung und eine hohe Intensität erzielen. Die ältere Anordnung nach HOSEMANN (3) benützt zu diesem Zweck ein Spaltsystem, die neuere nach KRATKY (4) eine massive Blockkamera, die große Vorteile hinsichtlich besserer Auflösung und leichterer Justierung bietet. Der strichförmige Primärstrahl hat zur Folge, daß eine Verzerrung der Streukurve auftritt, die bei unorientierten Diagrammen besonders groß ist und als „Verschmierung" bezeichnet wird. Durch ein von GUINIER und FOURNET (5) zuerst angegebenes und dann durch KRATKY, KAHOVEC, POROD (6) sowie durch GEROLD (7) erweitertes Verfahren (sog. „Entschmierung") ist es aber möglich, diese Verzerrung wieder zu eliminieren und die originale Streukurve herzustellen. Bei orientierten Präparaten allerdings ist eine Entschmierung nicht notwendig. Der ganze Fragenkomplex ist in einer neueren Arbeit (8) eingehend dargestellt worden.

Für feinere Kleinwinkeluntersuchungen ist es unbedingt erforderlich, mit monochromatischer Röntgenstrahlung zu arbeiten. Am gebräuchlichsten ist die Cu-Kα-Linie mit einer Wellenlänge von 1,54 Å. Da die

Monochromatisierung durch Reflexion an einem Kristall einen bedeutenden Intensitätsverlust zur Folge hat, begnügt man sich oft mit dem Filter-differenzverfahren nach Ross oder für weniger anspruchsvolle Unter-suchungen mit einer bloßen Filterung (Nickelfilter für Cu-Kα-Strahlung), um wenigstens den größten Teil der Bremsstrahlung wegzunehmen. Die Registrierung erfolgt mit Hilfe des photographischen Films oder durch ein Zählrohr. Beide Methoden haben ihre Vor- und Nachteile. Bei der photographischen Aufnahme bekommt man einen unmittelbaren Über-blick über das ganze Kleinwinkeldiagramm und kann außerdem infolge der akkumulierenden Wirkung des Films durch Erhöhung der Belich-tungszeit auch sehr kleine Intensitäten vermessen. Andrerseits muß der Film noch photometriert und die Photometerkurve auf Schwärzung um-gezeichnet werden, wobei jede dieser Operationen einen Verlust an Ge-nauigkeit bedingt. Das Zählrohr hingegen hat den Vorteil der direkten Messung. Man kann die Stoßrate (Stoßzahl pro Minute) unmittelbar als Maß für die Intensität ansetzen. Die Messung muß aber punktweise und für jeden Punkt über einen längeren Zeitraum erfolgen, um die natürlichen statistischen Schwankungen auszugleichen. Die flächenhafte Vermessung eines Diagramms würde zu untragbaren Arbeitszeiten führen. Die An-wendung des Zählrohrs kommt daher nur für unorientierte Präparate oder für Aufnahmen mit strichförmigem Primärstrahl in Frage, wo die Intensität nur längs einer Linie gemessen wird.

Als Nullpunkt des Kleinwinkeldiagramms ist grundsätzlich der Schwerpunkt der Intensitätsverteilung des Primärstrahls in der Regi-strierebene zu betrachten. Als Abszisse der Streukurve erscheint der Abstand eines Meßpunktes vom Nullpunkt in Zentimeter innerhalb der Registrierebene. Der Unterschied zwischen dem Winkel und dem Tangens kann dabei unbedenklich vernachlässigt werden, da nur sehr kleine Winkel in Frage kommen. Die Intensität der Streukurve kann in will-kürlichen Einheiten angegeben werden.

Bei der theoretischen Behandlung und Auswertung der KWS erweist es sich als zweckmäßig, die Unterscheidung zwischen verdünnten und dichtgepackten Systemen einzuführen, wie besonders von Kratky (9) betont worden ist. Es handelt sich um die Frage, ob die Interferenz der von verschiedenen Teilchen ausgehenden Sekundärwellen (interpartikuläre Interferenz) eine merkliche Rolle spielt oder nicht. Die Frage ist gleichwertig mit der anderen, ob zwischen den Partikeln regelmäßige Phasenbeziehungen bestehen. Man sieht unmittelbar ein, daß dies um so weniger der Fall sein wird, je weiter die Teilchen von einander entfernt und je unregelmäßiger sie angeordnet sind. In einer verdünnten Lösung können die interpartikulären Interferenzen meist vernachlässigt werden. Die KWS besteht dann nur aus der Summe der von den einzelnen Teil-chen herrührenden Beiträge (Partikelstreuung). Die Verdünnung ist aber hierfür nicht der einzige ausschlaggebende Faktor. Auch die Polydispersität [HOSEMANN (3, 10)] sowie eine stark anisometrische Teilchenform [KRATKY und POROD (11)] begünstigen das Auftreten einer reinen Partikelstreuung.

2. Verdünnte Systeme

Die theoretische Behandlung der verdünnten Systeme ist bereits zu einem recht befriedigenden Stand gediehen. Für die festen Hochpolymeren können diese detaillierten Auswertungsmethoden aber naturgemäß nur eine beschränkte Bedeutung haben. Denn es liegen nur selten wohldefinierte Teilchen vor und der Streueffekt wird im wesentlichen durch die interpartikulären Interferenzen bestimmt. Eine Auswertung als verdünntes System wird meist nur im hochgequollenen Zustand in Frage kommen und überschlagsmäßige Resultate liefern. Wir können uns daher im folgenden auf die wichtigsten Tatsachen beschränken.

Zunächst ist festzustellen, daß die Auswertung nur dann detaillierte Angaben liefern kann, wenn Teilchen einheitlicher Größe und Gestalt vorliegen, wie dies etwa bei den Eiweißstoffen der Fall ist. Bei polydispersen Systemen erhält man nur eine mittlere Größe und evtl. eine rohe Charakterisierung der Form sowie der statistischen Größenverteilung.

a) Auswertung nach Größe und Form

Wie GUINIER (12) in einer grundlegenden Arbeit gezeigt hat, kann die Streukurve von corpuscularen Teilchen, die nicht zu sehr von der Kugelgestalt abweichen, stets in guter Näherung durch eine Gaußsche Glockenkurve approximiert werden, in die die wahre Streukurve mit der Annäherung an den Winkel 0 asymptotisch übergeht:

$$I \sim M \cdot e^{-s^2 R^2/3}, \tag{1}$$

worin I die gestreute Intensität, M das Molekulargewicht und R den sog. Streumassenradius bedeuten. Letzterer ist so definiert, daß R^2 gleich dem Mittelwert der Quadrate aller vom Schwerpunkt zu den Elektronen gezogenen Abstände bedeutet. Er stellt mithin ein Maß für die mittlere Größe eines Teilchens dar. Um eine anschauliche Vorstellung zu geben, sei erwähnt, daß der Streumaßenradius einer homogenen Kugel das $\sqrt{3/5}$ fache des Kugelradius beträgt.

Der Streuwinkel θ erscheint in allen theoretischen Formeln implizit in der Variablen s, die gleichzeitig der experimentellen Abszisse m direkt und der Wellenlänge λ umgekehrt proportional ist:

$$s = \frac{4\pi}{\lambda} \sin \frac{\theta}{2} \doteq 2\pi\theta/\lambda \doteq 2\pi m/\lambda a p. \tag{2}$$

Hierin ist a der Abstand Präparat-Registrierebene und p das Übersetzungsverhältnis, mit dem die Streukurve graphisch aufgetragen wird.

Der Unterschied zwischen Sinus und Winkel kann wieder vernachlässigt werden.

Die Intensität beim Winkel 0 ist c.p. dem Molekulargewicht bzw. dem
Teilchenvolumen proportional. Dort treten ja keine Phasendifferenzen
auf, die Amplituden der Sekundärwellen addieren sich daher mit dem
vollen Betrage, so daß die resultierende Amplitude der Zahl der in einem
Teilchen wirksamen Elektronen entspricht. Die von *einer* Partikel ab-
gebeugte Intensität ist somit dem *Quadrat* der Elektronenzahl bzw. des
Molekulargewichtes proportional. Da aber die Zahl der Teilchen bei vor-
gegebener Substanzmenge im umgekehrten Verhältnis zur Größe steht,

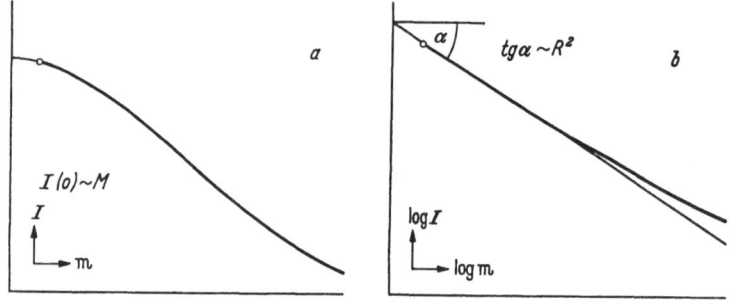

Abb. 1. Winkelabhängigkeit der Kleinwinkelstreuung corpuscularer Teilchen (schematisch). *a*) Streukurve.
b) Guinier-Auftragung

ergibt sich für das ganze System doch nur eine einfache Proportionalität
der Intensität mit dem Molekulargewicht.

Die Auswertung nach Formel (1) ist einfach. Man trägt den Log-
arithmus der Intensität gegen das Quadrat der Abszisse auf, wie in
Abb. 1 schematisch gezeigt ist. Die Kurve soll gegen den Nullpunkt hin
asymptotisch in eine Gerade übergehen. Die Neigungstangente ist dem
Quadrat des Streumassenradius proportional und dieser kann daraus im
Sinne von (1) und (2) leicht berechnet werden, womit bereits eine erste
Aussage über die Teilchengröße gewonnen ist. Im Falle eines polydispersen Systems erhält man das Z-Mittel von R^2, es machen sich also
ganz überwiegend die großen Teilchen bemerkbar.

Die Guinier-Auftragung eignet sich auch vorzüglich, um die Streukurve bis zum Winkel 0 zu extrapolieren. Damit ist die Möglichkeit
gegeben, auch das Molekulargewicht (bei polydispersen Systemen das
Gewichtsmittel desselben) aus der KWS zu bestimmen. Dazu muß allerdings die Absolutintensität, d. h. die wahre Relation der KWS zur
integralen Intensität des Primärstrahls bekannt sein und diese For-
derung bringt große experimentelle Schwierigkeiten mit sich. Außerdem
ist die Messung der sonstigen intensitätsbestimmenden Faktoren nur mit
geringer Genauigkeit möglich. Daraus erklärt sich auch, daß von dieser

naheliegenden Methode zur Molekulargewichtsbestimmung bisher nur wenig Gebrauch gemacht worden ist.

Wie vom Referenten (*13*) gezeigt wurde, läßt sich aber die schwierige Bestimmung der Absolutintensität umgehen, indem man eine integrale Größe, die später näher zu besprechende Invariante Q, einführt. Man bekommt so zwar nicht das Molekulargewicht selbst, aber als gleichwertige Größe das Teilchenvolumen V:

$$V = \frac{I(0)\,(\lambda\,a\,p)^3}{Q\,4\,\pi} \quad \text{mit} \quad Q = \int_0^\infty I(m)\cdot m^2\cdot dm\,. \quad (3)$$

Einen ersten qualitativen Hinweis auf die Teilchenform erhält man bereits aus der Betrachtung der Streukurve in der Guinier-Auftragung. Die Kurve weicht bei größeren Winkeln von der Geraden ab, und zwar bei einer exakten Kugel etwas nach unten. Für abweichende Teilchengestalten liegt die Kurve höher und erreicht etwa bei einer Dissymetrie 1:2 die optimale Anpassung an die Guinier-Gerade. Bei noch größerer Dissymetrie verläuft die Kurve bei nicht zu kleinen Winkeln oberhalb der Geraden. Genau dasselbe Verhalten der Streukurve wird auch durch eine etwaige Polydispersität verursacht, die daher vom Einfluß der Teilchenform nicht getrennt werden kann. Dies gilt grundsätzlich nicht nur bei dieser Methode sondern überall.

Eine weitere Möglichkeit zur Abschätzung der Form (bzw. der Polydispersität) besteht in dem Vergleich von Streumassenradius und Teilchenvolumen (*14*). Es gilt nämlich die leicht einzusehende Beziehung, daß von allen homogenen Körpern die Kugel bei gegebenem Volumen den kleinsten möglichen Streumassenradius, und bei gegebenem Streumassenradius das größte Volumen besitzt. Wenn wir aus dem nach (3) bestimmten Volumen V einen fiktiven Kugel-Streumassenradius R_0 berechnen, dann kann der Quotient R/R_0 nur größer als, bestenfalls gleich 1 sein. Er hat die Bedeutung eines Dissymetriefaktors und ist auch seiner Definition nach dem bekannten Dissymetriefaktor der Ultrazentrifuge analog. Ein praktischer Unterschied besteht nur hinsichtlich des Einflusses der Solvatation. Eine äußere Solvathülle ist für die Röntgenmethode unsichtbar, solange nicht eine beträchtliche Verdichtung des Lösungsmittels vorliegt. Bei einer inneren Solvatation kommt es auf die Feinheit der Verteilung der Solvatmoleküle an. Eine homogen gequollene Kugel zeigt den Dissymetriefaktor = 1. Er steigt um so mehr an, je größer sich das Lösungsmittel in Inseln verteilt. In diesem Fall wird allerdings auch die Bedeutung des röntgenographisch bestimmten Volumens unsicher, da Formel (3) unter der Voraussetzung eines homogenen Körpers abgeleitet ist.

Bei extrem anisometrischen Teilchen, Stäbchen oder Blättchen, versagt die Auswertung nach GUINIER. Zwar ist auch dann noch (1) als asymptotische Formel korrekt, der Gültigkeitsbereich liegt aber bei so kleinen Winkeln, daß er in die experimentell nicht zugängliche tote Zone fällt. Wie der Referent (15) zeigen konnte, läßt sich in diesem Extremfall doch wieder eine einfache Auswertung durchführen. Die Streukurve kann dann — abgesehen von den kleinsten, experimentell nicht erfaßbaren Winkeln — als das Produkt eines sog. Lorentzfaktors und einer Funktion aufgefaßt werden, die sich nur auf den Querschnitt bei Stäbchen bzw. auf die Dicke bei Blättchen bezieht. Der Lorentzfaktor beträgt in diesen beiden Fällen s^{-1} bzw. s^{-2}. Er kann daher einfach eliminiert werden, indem man die Streukurve mit der Abszisse bzw. dem Quadrat derselben punktweise durchmultipliziert. Die so erhaltene Querschnitts- oder Dickenstreukurve ist näherungsweise wieder durch eine Gaußsche Glockenkurve wie in (1) gegeben, mit der einzigen Änderung, daß der Koeffizient 3 durch 2 bzw. 1 ersetzt wird. Der Streumassenradius bezieht sich dann natürlich auf den Querschnitt oder die Dicke allein. Ihr Betrag folgt aus dem auf den Nullpunkt extrapolierten Wert in analoger Weise wie das Volumen corpuscularer Teilchen nach (3).

Obwohl die obigen Beziehungen streng nur für Teilchen abgeleitet sind, die in einer oder zwei Dimensionen unendlich groß sind, hat es sich in der Praxis herausgestellt, daß sie bereits bei durchaus endlichen Abmessungen eine sehr befriedigende Näherung darstellen. Man bekommt schon bei Stäbchen mit einem Achsenverhältnis 1:4 recht gute Resultate. Gerade bei den festen Hochpolymeren hat diese Auswertungsmethode mehr Anwendungsmöglichkeiten als die für corpusculare Teilchen. Ebenso werden exaktere Formbestimmungen kaum in Frage kommen, obwohl heute schon eine ganze Anzahl von streng berechneten Streukurven für verschiedene Körperklassen vorliegen (15, 16).

b) Auswertung nach Polydispersität

Wie bereits erwähnt, kann jede Streukurve auch durch eine bestimmte Polydispersität anstelle einer Teilchenform interpretiert werden. Dieser Gesichtspunkt ist vor allem von Hosemann (3, 10) in den Vordergrund gestellt worden. Um in diesem Sinne überhaupt zu einer Auswertungsmethode zu kommen, muß vorausgesetzt werden, daß alle Teilchen etwa gleiche Gestalt besitzen. In der Hosemannschen Theorie ist angenommen, daß es sich um ungefähr kugelförmige Teilchen handelt, deren Streukurven hinreichend durch die Guiniersche Näherung (1) beschrieben werden können. Die statistische Verteilung der Größen wird in Form einer Funktion vom Maxwell-Typ mit variablem Exponenten angenommen. Dieser Ansatz ist sehr anpassungsfähig und dürfte für die meisten Fälle ausreichen. Hosemann gelangt dann nach längerer Rechnung zu einer sehr einfachen Auswertungsmethode wie folgt:

Man multipliziere die Streukurve mit dem Quadrat der Abszisse. Es entsteht dann eine Kurve mit einem Maximum. Nun lege man an den äußeren Ast die Wendetangente an und bestimme die Lage b des Schnittpunktes mit der Abszissenachse, desgleichen die Abszisse a des Maximums. Der Grad der Polydispersität, d. h. die mittlere relative Schwankung der Teilchengrößen ist dann durch das sog. Ästeverhältnis bestimmt:

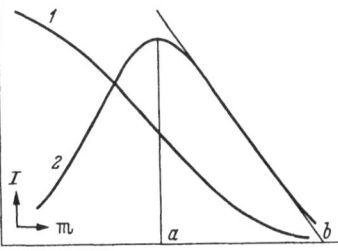

$$\left.\begin{aligned}\frac{b}{2\,a} - 1 &= 1 \,/\, \sqrt{2\,(n+1)} \\[2mm] \text{mit} \quad & \\[2mm] H(R) &= \frac{2}{R_0^{n+1}\left(\dfrac{n-1}{2}!\right)}\, R^n \cdot e^{-R^2/R_0^2},\end{aligned}\right\} \tag{4}$$

Abb. 2. Auswertung der Polydispersität nach HOSEMANN. 1... Streukurve; 2... dieselbe mit dem Quadrat der Abszisse multipliziert

wobei der Exponent n die Schärfe der statistischen Verteilung $H(R)$ bestimmt. Das arithmetische Mittel des Streumassenradius ergibt sich zu:

$$\overline{R} = \frac{1}{a}\, \sqrt{\frac{6}{n+2}} \cdot \left(\frac{n}{2}\right)! \Big/ \left(\frac{n-1}{2}\right)! \tag{5}$$

In der obigen Form sind die Gleichungen der Monographie von GUINIER und FOURNET (17) entnommen. HOSEMANN (10) selbst hat anstelle des Streumassenradius den Kugelradius verwendet; doch ist die Beschränkung auf eine genau definierte Teilchenform nicht notwendig, da ja doch mit der Guinierschen Näherung gerechnet wird und die Kugel nicht einmal die optimale Anpassung an diese gibt.

Ein gleichwertiges Verfahren wurde von SHULL und ROESS (18) angegeben, das aber in der Durchführung umständlicher ist und daher keinen Vorteil bietet. Übrigens ist auch eine Übertragung der obigen Formeln auf Stäbchen mit statistischer Verteilung der Querschnitte durchaus möglich, ebenso auf orientierte Ellipsoide, was durch HOSEMANN (3) ebenfalls geschehen ist.

HOSEMANN hat immer den Standpunkt vertreten, daß seine Auswertungsmethode, die ja mit der Voraussetzung einer reinen Partikelstreuung arbeitet, auch für relativ dichtgepackte Systeme zuständig ist, wenn nur die Polydispersität genügend hoch ist. Durch eine sehr allgemeine Beweisführung gelangte er zu folgendem Kriterium (10):

„Eine Auswertung als verdünntes System ist immer dann statthaft, wenn die relative statistische Schwankung der Teilchengrößen die Packungsdichte (= Volumfraktion) übersteigt."

Das Hosemannsche Kriterium gilt natürlich nur für ein im eigentlichen Sinn polydisperses System (in dem in allen Volumbereichen dieselbe statistische Größenverteilung vorliegt). Sobald eine Entmischung einsetzt, werden wir interpartikuläre Interferenzen erwarten müssen. Es muß daher durch andere Kriterien, als die Röntgenstreuung sie liefert, geprüft werden, ob das System wirklich polydispers ist.

3. Dichtgepackte Systeme

Bei dichter Packung der Teilchen werden im allgemeinen interpartikuläre Interferenzen auftreten. Wir können ihr Zustandekommen auf drei Effekte zurückführen (15b):

a) Undurchdringlichkeit. Wie Debye (19) in einer grundlegenden Arbeit gezeigt hat, bewirkt allein der Umstand, daß zwei Teilchen sich nicht weiter nähern können als auf die Distanz ihres Durchmessers, auch ohne das Wirken von besonderen ordnenden Kräften eine Art von verschwommener Interferenz in der Streukurve. Durch Polydispersität wird der Effekt herabgesetzt (10, 20, 21), doch gehen die Meinungen über die Größe dieses Einflusses auseinander.

b) Assoziation. Eine Zusammenlagerung von Teilchen zu größeren Aggregaten (Cluster) ist ebenfalls häufig anzutreffen. Sie bewirkt einen besonders steilen Anstieg der Streukurve bei kleinsten Winkeln.

c) Ordnung. Sie kann durch starke Kräfte zwischen den Teilchen (vor allem in Lösungen) oder in Festkörpern durch die Entstehungsbzw. Wachstumsbedingungen hervorgerufen werden. Der Einfluß einer regelmäßigen Anordnung kann die Streukurve so stark modifizieren, daß von einer Partikelstreuung nichts mehr übrig bleibt.

Die vollständige Beschreibung eines dichtgepackten Systems würde die Angabe von Größe und Form der kolloiden Teilchen mit eventueller Polydispersität, sowie ihrer gegenseitigen Lage und Orientierung im Raum erfordern. Eine solche erschöpfende Auskunft kann die KWS nicht liefern, jedenfalls nicht in Form einer allgemein anwendbaren Methode. Der naheliegendste Weg, für verschiedene Modelle die Streukurven zu berechnen und mit den experimentellen zu vergleichen, scheitert an der unendlichen Vielzahl der möglichen Systeme und an der Schwierigkeit der theoretischen Behandlung. Eine solche modellmäßige Auswertung ist daher nur in besonderen Fällen möglich. An dieser Stelle wären auch die von Kratky (9), sowie vor allem von Hosemann u. Mitarb. (22) durchgeführten lichtoptischen Modellversuche zu erwähnen, die eine aussichtsreiche Möglichkeit bieten, um die schwierige Rechnung zu umgehen. In vielen Fällen wird man sich aber doch damit begnügen müssen, unter Verzicht auf eine vollständige Beschreibung des Systems nur gewisse Parameter oder Kennzahlen aus der KWS abzuleiten.

Für die Auswertung macht es einen großen Unterschied, ob das System und damit auch die KWS — im großen betrachtet — isotrop ist, oder ob eine Orientierung vorliegt. Außerdem ist zu unterscheiden zwischen dem Fall, daß die interpartikulären Interferenzen nur eine Modifikation der Partikelstreuung bewirken, und dem anderen, daß sie überhaupt ausschlaggebend sind. Die Streukurve ist im ersten Falle monoton fallend (diffus), im zweiten zeigt sie Interferenzmaxima (diskret). Damit ist eine natürliche Einteilung gegeben, nach der wir uns im folgenden richten wollen.

a) Unorientierte Systeme, diffuse KWS

Die allgemeine Theorie, die von DEBYE und BUECHE (*23*), sowie vom Referenten (*13*) entwickelt wurde, geht von der Tatsache aus, daß für die KWS nur die Schwankungen der Elektronendichte maßgebend sind. Die Intensität ist bestimmt durch das mittlere Schwankungsquadrat $\overline{\eta^2}$, das daher auch als Streukraft des Systems bezeichnet wird. Die Winkelabhängigkeit der KWS, also die Streukurve, hängt zusammen mit der räumlichen Verteilung der Elektronendichte, die durch die sog. Korrelationsfunktion („Charakteristik" des Referenten) beschrieben wird. Ohne auf Einzelheiten einzugehen sei nur erwähnt, daß die Korrelationsfunktion in ähnlicher Weise wie die Pattersonfunktion der Kristallstrukturanalyse definiert ist. Es ist also die Verteilung für jedes Punktepaar zu berücksichtigen. Die Streukurve hängt mit der Korrelationsfunktion durch eine Fouriertransformation zusammen und kann prinzipiell durch eine Fourierumkehr gewonnen werden. Daraus folgt, daß aus der KWS nur solche Informationen erhalten werden können, die in irgendeiner Form in der Korrelationsfunktion enthalten sind.

Es sei besonders betont, daß diese Betrachtungsweise keinen Gebrauch von der Vorstellung diskreter Teilchen macht. Für die Korrelationsfunktion ist es gleichgültig, ob zwei Punkte innerhalb ein und desselben Teilchens oder in zwei verschiedenen liegen. Die interpartikulären Interferenzen sind daher bereits implizit in der Theorie berücksichtigt.

Als erster allgemeiner Parameter sei nun die Streukraft behandelt. Wenn ein System aus zwei Phasen mit den Volumanteilen w_1 und w_2 sowie den respektiven Elektronendichten ϱ_1 und ϱ_2 vorliegt, folgt aus der Definition:

$$\overline{\eta^2} = \overline{(\varrho_1 - \varrho_2)^2}\, w_1 w_2 . \tag{6}$$

Ebenso läßt sich leicht der Ausdruck für ein System, bestehend aus beliebig vielen Phasen, angeben, wenn nur alle Volumfraktionen und Elektronendichten bekannt sind. Umgekehrt kann die aus der KWS bestimmte Streukraft dazu benützt werden, um zu Aussagen über die Zusammensetzung des Systems zu gelangen. Zwar ist ein solcher Rückschluß natürlich nicht eindeutig, es kann aber doch eine Entscheidung zwischen mehreren zur Diskussion stehenden Möglichkeiten getroffen werden. Die Streukraft eignet sich daher zur Untersuchung der Frage

25*

der „kristallinen" und „amorphen" Bereiche sowie der Hohlräume in festen Hochpolymeren.

Zur experimentellen Bestimmung kann nicht die Intensität an irgend-einer Stelle genommen werden, weil diese ja winkelabhängig ist, worin zum Ausdruck kommt, daß sie nicht nur vom Betrage der Elektronen-dichtenschwankungen, sondern auch von der räumlichen Verteilung der-selben abhängt. Man braucht vielmehr eine integrale Größe, die sog. Invariante (13), die gegenüber der geometrischen Struktur unveränder-lich ist:

$$Q = \int_0^\infty I(m)\, m^2 \cdot dm \; ; \quad \widetilde{Q} = \int_0^\infty \widetilde{I}(m)\, m \cdot dm = 2\, Q \; ; \quad \overline{\eta^2} \sim Q \, . \quad (7)$$

Die Invariante kann demnach ebenso aus der unverschmierten wie aus der verschmierten Kurve gewonnen werden, wobei die auf den letzteren Fall bezogenen Werte hier durch eine hochgestellte Tilde gekennzeichnet sind. Die Invarianz gegenüber der geometrischen Verteilung ist eine ein-fache Folge der Fouriertransformation, wie auch aus der Theorie von Debye und Bueche sowie aus den Darlegungen von Guinier und Fournet in ihrer Monographie (17) hervorgeht. Man kann diese Eigen-schaft aber auch, zumindest im Hinblick auf die Größe der Inhomogeni-täten, leicht unmittelbar einsehen. Stellen wir uns nämlich vor, daß ein kolloides System im linearen Verhältnis $1:p$ vergrößert wird, dann müßte die Intensität mit p^3 steigen, die Abszisse m aber auf $1/p$ des entsprechen-den ursprünglichen Wertes sinken. Da m im ersten Integral in (7) in der dritten Potenz vorkommt, kompensieren sich beide Änderungen genau.

Die Invariante Q bzw. $\widetilde{Q}$ stellt bereits ein Maß für die Streukraft dar. Um diese aber in absoluten Einheiten zu erhalten, ist eine Umrechnung erforderlich, in die die integrale Intensität des Primärstrahls, die Geome-trie der Kamera und die Abmessungen des Präparates eingehen, und die zu umständlich ist, um an dieser Stelle im einzelnen angeführt zu werden.

Als weiterer Parameter kann aus der KWS ein Maß für die mittlere ·Größe der Inhomogenitäten gewonnen werden, und zwar in verschiedener Weise. Zunächst ist zu sagen, daß auch bei dichtgepackten Systemen für das asymptotische Verhalten bei kleinsten Winkeln eine ähnliche Beziehung gilt wie die Guiniersche Näherung bei reiner Partikelstreuung. Man kann zwar im allgemeinen keine Gaußsche Glockenkurve erwarten, aber grundsätzlich muß die Streukurve sich im zweiten Grade anschmie-gen. Mathematisch kommt das so zum Ausdruck, daß die Streukurve, als Potenzreihe angeschrieben, immer eine gerade Funktion ist. Man kann die Auswertung wie bei den verdünnten Systemen führen und eine dem Streumassenradius analoge Größe gewinnen. Von Debye und Bueche (23) wurde das $\sqrt{2}$-fache derselben als Korrelationslänge bezeichnet. Wir können diesen Parameter auffassen als den Streumassenradius eines kugelsymmetrischen Ersatzkörpers, dessen Elektronendichte durch die Korrelationsfunktion gegeben ist. Man kann daher die Korrelationslänge durchaus als ein anschauliches Maß für die Größe der Inhomogenitäten betrachten.

Vom Referenten wurde die sog. Kohärenzlänge (13) eingeführt, die als die integrale Breite der Korrelationsfunktion definiert ist. Sie hat gleichzeitig die anschaulichere Bedeutung einer mittleren Durchschußlänge. Damit ist folgendes gemeint. Wir stellen uns vor, daß die Teilchen durch Strahlen in jeder Lage und Richtung durchsetzt werden. Die dabei herausgeschnittenen Teilstücke stellen die Gesamtheit der überhaupt vorkommenden Abmessungen dar und ein Mittelwert derselben repräsentiert offenbar ein Maß für die mittlere Größe. Die Kohärenzlänge ist nun einfach das gewogene Mittel aller dieser Durchschußlängen. Die experimentelle Bestimmung ist sehr einfach. Die Kohärenzlänge ist nämlich der leicht bestimmbaren integralen gestreuten Intensität proportional. Die Auswertung kann übrigens ohne weiteres mit der spaltverschmierten Streukurve vorgenommen werden. Für eine Kugel beträgt die Kohärenzlänge $3/4$ des Durchmessers.

Eine Auswertung nach Molekulargewicht oder Volumen der kolloiden Teilchen, wie bei den verdünnten Systemen besprochen, kommt für dichtgepackte Systeme nicht in Frage. Die auf den Winkel 0 extrapolierte Intensität ist hier in der gleichen Weise mit den Dichteschwankungen verknüpft wie bei der Streuung des sichtbaren Lichtes. Es handelt sich dabei nicht um die Schwankungen der Elektronendichte an sich, die ja in der Streukraft ihren Ausdruck finden, sondern um die Unterschiede zwischen den über größere Teilbereiche gebildeten Mittelwerten der Elektronendichte. Die Intensität verschwindet daher beim Winkel 0 nicht nur im Falle eines homogenen Systems, etwa einer reinen Flüssigkeit, sondern auch dann, wenn die Schwankungen der Elektronendichte zwar vorhanden sind, sich aber bereits innerhalb kleiner Bereiche kompensieren. Das würde z. B. dann der Fall sein, wenn sich Teilchen aus einer ursprünglich homogenen Matrix gebildet haben, wobei die Substanz aber nur aus der nächsten Umgebung zusammengezogen wurde. Auf diese Verhältnisse haben besonders GUINIER und BELBÉOCH (24) hingewiesen. Sie dürften mehr Beachtung verdienen, als ihnen bisher zuteil wurde.

Die bisher besprochenen Parameter waren ohne Bezugnahme auf eine spezielle Strukturvorstellung definiert. Eine weitere wichtige Gesetzmäßigkeit ergibt sich nun, wenn wir den Spezialfall eines Zweiphasensystems betrachten, wobei wir annehmen wollen, daß keine Ordnung über weitere Bereiche besteht. Wie vom Referenten (13) sowie von DEBYE (25) u. Mitarb. theoretisch abgeleitet werden konnte, sollte die Streukurve bei größeren Winkeln einen asymptotischen Auslauf $\sim s^{-4}$, bzw. im Falle der spaltverschmierten Streukurve $\sim s^{-3}$ erreichen, wobei die Intensität dieses Kurvenastes der inneren Oberfläche proportional ist. Dieselbe Beziehung wurde von VAN NORDSTRAND und HACH (26) auf rein empirischem Wege gefunden. Die Auswertung ist im Prinzip einfach. Man multipliziert die Streukurve mit der vierten (bzw. die verschmierte Kurve mit der dritten) Potenz der Abszisse m. Falls die Kurve dann einem konstanten Grenzwert zustrebt, ist dieser der inneren Oberfläche proportional:

$$0_{sp} \sim \lim (m^4 I) \sim \lim (m^3 \tilde{I}) , \quad \text{für großes } m . \quad (8)$$

Die absolute Bestimmung der inneren Oberfläche kann entweder mit Hilfe der Absolutintensität oder der Invariante erfolgen. Doch dürfte es zweckmäßiger sein, nur Vergleichsmessungen relativ zu einem bekannten Eichpräparat durchzuführen.

Die Kleinwinkelmethode zur Ermittlung der inneren Oberfläche hat sich bisher hauptsächlich bei anorganischen Pulvern (Katalysatoren u. dgl.) bewährt. Soweit Absolutwerte bestimmt sind, scheinen sie in der Regel höher zu liegen als die nach der Adsorptionsmethode gewonnenen (25). Das mag daran liegen, daß durch die KWS auch noch solche Hohlräume erfaßt werden, die für das adsorbierte Gas unzugänglich sind. Bei den Hochpolymeren muß man aber auch mit einer Umkehrung des Verhältnisses rechnen, indem Bereiche mit molekularer Aufspaltung (Fransen, „amorphe" Bereiche) wohl in der Adsorption, nicht aber röntgenographisch in Erscheinung treten. Auch aus diesem Grunde dürfte es rätlich sein, mit Relativwerten zu arbeiten.

b) Orientierte Systeme, diffuse Streuung

Bei den festen Hochpolymeren hat man in der Regel mit einer Faserorientierung zu rechnen. Die KWS ist dann anisotrop, d. h. sie ist verschieden in Richtung des Meridians und des Äquators. Soweit es sich um *diffuse* Streuung handelt, hat bis jetzt nur die letztere eine Bedeutung. Man verwendet daher eine Spaltkamera und justiert die Faser parallel zum Strich des Primärstrahls. Normal dazu wird die äquatoriale Streuung vermessen.

Bei der Bestimmung der Invariante und der Streukraft ändert sich nichts. Man kann genau so vorgehen, wie für die spaltverschmierte Streukurve von unorientierten Präparaten beschrieben wurde. Der Grund liegt darin, daß es sich hier um eine integrale Größe handelt. Die Intensitätsverteilung in Richtung des Meridians, die ja nicht gesondert vermessen wird, ist bereits implizit mitberücksichtigt, da die Verwendung eines strichförmigen Primärstrahls gleichbedeutend ist mit einer Integration über die meridiale Streukurve.

Für alle anderen Größen ist aber die Tatsache der Orientierung von Bedeutung. Die äquatoriale Streukurve entspricht der Querschnittstreukurve von stäbchenförmigen Teilchen in verdünnten Systemen. Sie steht

mit der Elektronendichtenverteilung in einer Ebene normal zur Faserrichtung in einem analogen Zusammenhang wie die gewöhnliche Streukurve unorientierter Systeme mit der Elektronendichtenverteilung im
Raum. Korrelationslänge und Kohärenzlänge können zwar ebenfalls bestimmt werden, beziehen sich aber nur auf die Äquatorebene. Desgleichen
könnte man das Verfahren zur Messung der inneren Oberfläche auf
orientierte Systeme übertragen und würde dann die Länge der Grenz-
linie in der Ebene erhalten. Doch dürfte dies keine Bedeutung besitzen.

Aus dem Gesagten geht hervor, daß es manchmal erwünscht sein
kann, die Orientierung wieder aufzuheben, um die normalen Auswertungsmethoden anwenden zu können. Eine Zerkleinerung des Präparates und
Einfüllen in eine Capillare führt nur bedingt zum Ziel, weil sich die
kleinen Faserstückchen doch meist wieder irgendwie ordnen. Man kann
aber den Rest von Orientierung beseitigen, indem man das Präparat
einfach dreht, und zwar genügt es, dies von Hand aus zu tun. Es mag
hier angebracht sein, darauf hinzuweisen, daß es für die Frage: orientiert
oder unorientiert, nur darauf ankommt, daß das Kleinwinkeldiagramm
isotrop erscheint. Daß im Innern des Präparates über makroskopische
Bereiche hinweg nach wie vor eine Orientierung besteht, ist dabei
gleichgültig.

Es erhebt sich nun die Frage, in welchem Zusammenhang die Streukurve des isotrop gemachten Präparates mit der ursprünglichen äquato-
rialen Streukurve steht. Die Antwort ergibt sich aus dem, was über den
Lorentzfaktor von stäbchenförmigen Teilchen bei den verdünnten Systemen gesagt wurde. Die äquatoriale Streukurve sollte identisch sein
mit derjenigen Kurve, die erhalten wird, wenn man die Streukurve des
isotropen Systems entschmiert und dann mit der Abszisse multipliziert.
Das ist eine überprüfbare Aussage. Der Nachweis ihrer Richtigkeit wurde
durch KRATKY (27) und HERMANS (28) erbracht. Damit sind gleichzeitig
indirekt auch die anderen vorstehenden Behauptungen über die Auswertung orientierter Systeme verifiziert.

Wenn blättchen- oder lamellenförmige Teilchen vorliegen, gelingt es,
durch Walzen eine höhere Orientierung herzustellen, wie von BURGENI
und KRATKY (29) bei der Cellulose und von KRATKY und KURIYAMA (30)
an Seidenfibroin gezeigt wurde. Die Anisotropie im Querschnitt kann
bei Durchstrahlung in der Faserrichtung unmittelbar sichtbar gemacht
werden. Dazu ist allerdings die Verwendung einer Lochkamera erforderlich, was eine quantitative Auswertung sehr erschwert. Wenn man mit
einer Spaltkamera arbeitet, sind zwei Aufnahmen mit einer gegenseitigen
Verdrehung von 90° notwendig. Grundsätzlich ist es dann möglich, zu
Aussagen über beide Querschnittdimensionen zu gelangen. Da aber
exakte experimentelle Untersuchungen in dieser Richtung bisher nicht
vorliegen, wollen wir auf eine nähere Besprechung verzichten.

c) Geordnete Systeme, diskrete Streuung

Das Auftreten von diskreten Maxima kann als der Ausdruck einer periodischen oder zumindest quasiperiodischen Ordnung angesehen werden. Soweit es sich um scharfe Reflexe mit höheren Ordnungen handelt, wie sie etwa beim Kollagen von Kratky und Sekora (31) sowie von Bear (32) gefunden wurden, sind einfach die Methoden der Kristallstrukturanalyse zuständig, deren Behandlung hier zu weit führen würde. Die Kleinheit der Winkel macht aber die Verwendung der apparativen Hilfsmittel der KWS notwendig. Insbesondere werden an das Auflösungsvermögen der Kamera hohe Anforderungen gestellt.

Eine so ausgezeichnete gittermäßige Ordnung wie beim Kollagen ist eine Ausnahme. Die Reflexe, die man bei den festen Hochpolymeren im Kleinwinkelgebiet erhält, sind recht diffus und zeigen nur selten höhere Ordnungen. Außerdem handelt es sich meist nur um eine Periodizität in einer Richtung. Das alles macht eine gesonderte Besprechung notwendig, wobei die besondere Art der statistischen Störungen sowie die Kleinheit der geordneten Bereiche in ihrem Einfluß auf die KWS zu diskutieren sind.

Bei einer eindimensionalen quasiperiodischen Anordnung von Teilchen haben wir sog. Störungen zweiter Art zu erwarten, wie sie von Zernicke und Prins (33) sowie von Kratky (34) zur Diskussion der Flüssigkeitstruktur herangezogen worden. Damit ist gemeint, daß jedes Teilchen nicht wie in einem Kristall um eine fixierte Ruhelage, sondern relativ zu seinen nächsten Nachbarn schwankt. Die absoluten Abweichungen von der einer strengen Periodizität entsprechenden Lage nehmen daher mit der Entfernung von einem beliebig gewählten Bezugspunkt zu, und zwar mit der Wurzel aus der Entfernung. Für die Beugung an einem solchen linearen Gitter mit statistischen Schwankungen zweiter Art hat J. J. Hermans (35) eine ganz allgemeine mathematische Lösung angegeben. Speziellere Modelle wurden von Hosemann (36) sowie vom Referenten (21, 37) behandelt. Das Ergebnis dieser Rechnungen ist, daß man diffuse Maxima zu erwarten hat, deren Lage nur noch ungefähr nach dem Braggschen Gesetz durch die mittlere Periode bestimmt wird, wobei die Reflexe höherer Ordnung rasch breiter werden und zusammenfließen. Ab einer relativen statistischen Schwankung von etwa 25% ist praktisch nur noch die erste Ordnung zu erkennen. Falls das lineare Gitter aus blättchenförmigen Teilchen besteht, die der Dicke nach geordnet sind (sog. Micellenpaket), ergibt sich eine weitere Schwächung der Maxima durch den Lorentzfaktor, wie vom Referenten (37) gezeigt wurde. Bei genügend hoher Schwankung kann dann auch die erste Ordnung der Reflexe praktisch verschwinden.

Die Schärfe und Ausbildung der Reflexe hängt auch von der Größe der streuenden Bereiche ab. Der bekannten Linienbreite der Weitwinkeldiagramme entspricht im Kleinwinkelgebiet die Verbreiterung der Reflexe infolge geringer Zahl von sich periodisch wiederholenden Elementen. Doch überschneidet sich dieser Effekt mit der durch die Schwankungen der Periodizität bewirkten Verbreiterung und wird wohl kaum von dieser abgetrennt werden können.

Wichtiger und aufschlußreicher ist die seitliche Ausbildung der Reflexe. Eine periodische Punktreihe sollte Schichtlinienreflexe geben, d. h. die Intensität ist unbegrenzt auf einer Linie normal zur Periode auseinandergezogen. Je breiter die streuenden Elemente sind, um so kürzer werden die Schichtlinien, wie in Abb. 3 veranschaulicht ist. BOLDUAN und BEAR (38) haben diesen Zusammenhang für die Reflexe sehr des Kollagens eingehend diskutiert. Das Verständnis wird sehr erleichtert, wenn man die von EWALD (39) in die Kristallstrukturanalyse eingeführten Faltungsoperationen heranzieht. Ihre Bedeutung für die KWS ist vor allem von HOSEMANN (40) betont worden. Ohne mathematische Formeln läßt sich der Sachverhalt wie folgt ausdrücken: Die Intensitätsverteilung eines Reflexes um seinen Schwerpunkt ist identisch mit dem Kleinwinkeldiagramm, das der kristalline Bereich für sich allein geben würde. In diesem Sinne kann die KWS als der Reflex nullter Ordnung angesehen werden,

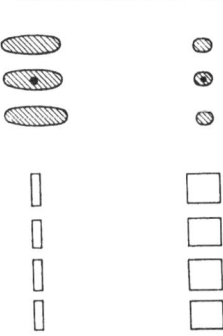

Abb. 3. Schema eines linearen Gitters mit verschiedenen Querdimensionen und entsprechende Kleinwinkelstreuung

wie auch in Abb. 3 zum Ausdruck kommt. Tatsächlich wird man kaum je eine solche Identität finden, weil zur eigentlichen KWS auch nicht gittermäßig geordnete Bereiche beitragen und weil die Verhältnisse in den Reflexen durch die Gitterstörungen modifiziert werden. Es bleibt aber jedenfalls bestehen, daß man die Intensitätsverteilung längs einer Schichtlinie wie eine gewöhnliche Streukurve hinsichtlich der Querdimension der „Kristallite" auswerten darf.

In Fällen, wo die Schichtlinienverbreiterung sich in irgendeiner Form mit der Ordnung ändert (z. B. keine meßbare Verbreiterung auf dem Äquator), ist diese Auswertung selbstverständlich nicht mehr zulässig. Es ist dann eine noch offene Frage, ob man die Beobachtungen anhand eines speziellen Modells klassisch interpretieren kann oder ob man allgemein auf das statistisch gestörte Modell von BONART und HOSEMANN (40 a) (gewellte Schichten) zurückgreifen soll.

Bisher war die Rede von einer eindimensionalen quasiperiodischen Ordnung. Die gleichzeitige Berücksichtigung einer Periodizität mit Störungen zweiter Art in seitlicher Richtung stellt ein sehr schwieriges mathematisches Problem dar, das in der Theorie des Parakristalls von HOSEMANN (40, 41) gelöst ist. Eine eingehende Darstellung ist hier nicht

möglich. Wir wollen uns daher damit begnügen, die bei einem Parakristall zu erwartenden Effekte qualitativ zu diskutieren. Nachdem der
Parakristall so definiert ist, daß er in jeder Gitterrichtung Störungen
zweiter Art aufweist, gilt auch hier für jeden Reflex das über das entsprechende lineare Gitter Gesagte. Dazu kommt aber noch, daß jetzt
die „Netzebenen" statistisch unregelmäßig verbogen und gewellt sind.
Das hat zur Folge, daß nur kleine Stücke der Netzebenen als eben erscheinen. Der Parakristall wirkt daher in seitlicher Richtung kleiner, als
er ist. Die Reflexe müssen entsprechend in der Schichtlinie verbreitert
sein. Zweitens sind die Netzebenen wegen ihrer Welligkeit auch effektiv
dicker. Der zugehörige Strukturfaktor fällt schneller ab und die höheren
Ordnungen des Reflexes werden schneller geschwächt. Alles in allem sind
also in einem Parakristall ähnliche Erscheinungen zu erwarten wie bei
einem linearen Gitter mit Störungen zweiter Art, nur daß die Reflexe
noch diffuser werden und ihre Zahl noch geringer ist.

II. Ergebnisse an festen Hochpolymeren
1. Unorientierte Präparate

Systematische Untersuchungen an unorientierten Hochpolymeren
sind bisher der Natur der Sache entsprechend hauptsächlich im hochgequollenen Zustand ausgeführt worden. Im Vordergrund des Interesses
stand dabei die Cellulose. KRATKY, JANESCHITZ und POROD (42) konnten
an luftgequollenen Regeneratcellulosefäden (Quellungsgrad bis zu 5,7)
den Fall einer praktisch reinen Partikelstreuung realisieren. Die Streukurve konnte interpretiert werden durch die Annahme von blättchenförmigen Teilchen (Micellen) mit schwankender Dicke. Durch Vergleich
mit den für verschiedene Ansätze berechneten Kurven ergab sich als
beste Anpassung eine mittlere Micelldicke von 50 Å mit einer relativen
Schwankung von 50%. Dieser Befund war in guter Übereinstimmung
mit den Ergebnissen, die in derselben Arbeit und früher von KRATKY (9)
an ungequollenen Fäden auf Grund der Theorie des Micellenpaketes,
also unter Berücksichtigung der interpartikulären Interferenzen erhalten
wurden, obwohl die Streukurven in beiden Fällen völlig verschieden sind.

Nun hat sich seither immer mehr die Überzeugung durchgesetzt, daß
die morphologische Struktur der Hochpolymeren viel regelmäßiger ist,
als früher angenommen wurde. Eine sehr hohe statistische Schwankung
ist nicht wahrscheinlich. Eine neuerliche Überprüfung ergab, daß die
Streukurve des luftgequollenen Systems auch durch die Annahme von
einheitlichen stäbchenförmigen Teilchen mit einem Querschnitt von
55×136 Å gedeutet werden kann. Diese Mehrdeutigkeit liegt im
Wesen der Methode. Jedenfalls aber bleibt der Wert für die Micelldicke
bestehen.

Auch an hoch-wassergequollenen Regeneratcellulosefäden (*43*) wurde
eine Partikelstreuung realisiert, die in analoger Auswertung eine Micelldicke von 31 Å ergab. Die nächstliegende Erklärung der Diskrepanz
gegenüber dem früheren Wert ist, daß in dem einen Falle nur der
„kristalline Kern", im andern aber das ganze Teilchen samt „Rinden-
schicht" vorliegt.

Native Cellulose kann nicht so hochgequollen werden, um eine Auswertung als verdünntes System zu rechtfertigen. Doch liegt eine Untersuchung von KRATKY und SEMBACH (*43*) an einem Rånbyschen „Micell-
pulver" vor, das durch hydrolytischen Abbau von Baumwolle erhalten
worden war. In wäßriger Aufschlämmung liefert es eine Partikelstreuung,
die durch Stäbchen mit einem Querschnitt von 32 × 93 Å interpretiert
werden konnte. Das ist praktisch dasselbe Ergebnis, wie bei der hoch-
wassergequollenen Regeneratcellulose gefunden wurden. Wiederum über-
einstimmende Werte wurden bei sog. Signerfäden erhalten (*43*). Diese
werden über eine extrem verdünnte Viscoselösung hergestellt, so daß der
Verdacht ausscheidet, es könnten Teilchen der nativen Cellulose erhalten
geblieben sein.

Von anderen Hochpolymeren liegen nur gelegentliche Kleinwinkel-
beobachtungen an unorientierten Systemen vor, die hier nicht gesondert
anzuführen sind. Dagen müssen die Arbeiten von KRATKY (*44*) u. Mitarb.
über das Seidengel erwähnt werden. Dieses liefert im gequollenen Zustand
ein besonders eindrucksvolles Beispiel für eine Partikelstreuung von
stäbchenförmigen Teilchen. Nach Eliminierung des Lorentzfaktors ergibt
sich eine Querschnittstreukurve vom Gaußchen Typ. Unter der Annahme
von einheitlichen Teilchen konnten die Autoren einen Querschnitt von
48 × 96 Å bestimmen. Ein damit praktisch übereinstimmender Betrag
für die Querschnittsfläche wurde aus der auf den Winkel 0 extrapolierten
Intensität erhalten. Diese Auswertung ist natürlich mit dem Vorbehalt
der Voraussetzung eines homodispersen Systems zu verstehen. Im Falle
von Polydispersität würden sich andere Dimensionen ergeben, eine eindeutige Aussage wäre aber nicht möglich. Übrigens darf an dieser Stelle
daran erinnert werden, daß die oben erwähnte Übereinstimmung der
Werte nach zwei verschiedenen Methoden keinen Beweis für die Richtigkeit der Grundannahme darstellt. Es gilt vielmehr immer, daß streuungs-
äquivalente Systeme in allen allgemein aus der KWS bestimmbaren
Parametern übereinstimmen müssen. Auch bei einem polydispersen System würde daher die aus der Intensität beim Winkel 0 berechnete Fläche
mit dem Gewichtsmittel aus den Querschnitten zusammenfallen, die aus
der Form der Streukurve erhalten werden.

Eine ältere Arbeit von MACARTHUR und PATNAIK (*45*) über Alginat-
fäden genügt zwar in der quantitativen Auswertung nicht allen An-
sprüchen, ist aber in den qualitativen Ergebnissen bemerkenswert. Es

ergab sich nämlich ein auffallender Unterschied je nach dem verwendeten Quellungsmittel. Mit Natrium- sowie mit Rubidiumcarbonat wurde überhaupt keine vermeßbare KWS erhalten, eine sehr intensive hingegen bei Ca··-gequollenen Fäden. Die Guinier-Auftragung zeigte deutlich eine geknickte Gerade, was am einfachsten durch das Vorliegen von zwei verschiedenen Teilchengrößen gedeutet werden kann. Die Autoren geben Fibrillendurchmesser von 135 und 177 Å an. Wenn auch diese Zahlen vielleicht anfechtbar sind, wird doch durch die Kleinwinkelmethode klar demonstriert, daß eine Aufspaltung in diskrete Partikel nur mit Ca··-Ionen eintritt, während mit Alkali-Ionen offenbar die Gelstruktur erhalten bleibt, deren Aggregate zu groß sind, um eine vermeßbare KWS zu liefern.

2. Orientierte Systeme, meridiale Kleinwinkelstreuung

a) Meridianreflexe

1944 machten Hess und Kiessig (46) die überraschende Entdeckung, daß im Kleinwinkeldiagramm von synthetischen hochpolymeren Faserstoffen (Polyamide, Polyester) am Meridian, d. h. in Richtung der Faserachse, deutliche Reflexe auftreten, die nach dem Braggschen Gesetz einer großen Periode von 70 bis 200 Å entsprechen. Seitdem sind durch die Arbeiten von Hess und Kiessig (2, 47); Arnett, Meibohm und Smith (48); Fankuchen und Mark (49); Zahn (50); Rothe (51) und anderen diese Befunde wiederholt bestätigt und auf weitere Hochpolymere ausgedehnt worden. Ja sogar an Glasfäden wurden von Statton (52) ähnliche Effekte beobachtet. Es handelt sich also um eine ganz allgemeine Erscheinung. Nur bei wenigen von den bekannteren synthetischen Hochpolymeren wie Teflon, Polystyrol und Orlon sind bisher keine Meridianreflexe mit großen Perioden gefunden worden. Zu Orlon ist jedoch zu bemerken, das Kratky, Sekora und Breiner (53) an Vinyon-N, einem Mischpolymerisat von Vinylchlorid und Acrylnitril, die Andeutung eines Reflexes entsprechend 170—180 Å beobachtet haben. Er tritt allerdings erst nach geeigneter Vorbehandlung auf.

Eine gewisse Sonderstellung nimmt in diesem Zusammenhang die Cellulose ein. Wenn man von einer unbestätigt gebliebenen älteren Beobachtung von Clark und Parker (54) absieht, waren Langperioden zunächst nicht bekannt. Nun besteht ja ein wesentlicher Unterschied in den Herstellungsbedingungen, da Cellulosefäden aus einem Fällbad bei niedriger Temperatur und nicht wie die synthetischen Hochpolymeren aus der Schmelze gezogen werden. Ausgehend von der Überlegung, daß hierin wahrscheinlich das Fehlen der großen Periode begründet ist, versuchte Kiessig (2, 55) diese durch eine geeignete Behandlung nachträglich hervorzurufen. Tatsächlich erhielt er nach Erhitzen von Cellulosefäden in Wasser bei 200° C einen deutlichen Meridianreflex entsprechend

204 Å. Auch partielle Hydrolyse führte zum Erfolg, wobei die Langperioden je nach Behandlung Werte von 110 bis 180 Å zeigten. Schließlich konnte KIESSIG auch an einer unbehandelten Reifenseide einen allerdings nur schwachen Reflex mit etwa 150 Å feststellen. Ähnliche Beobachtungen, besonders auch an unbehandelten Fasern, liegen von STATTON (56) vor. Auch an jodierter Kunstseide wurden Werte von 125 und 250 Å durch KRATKY und SEKORA (57) gefunden. Es spricht mithin alles für die Vermutung von KIESSIG, daß auch in der regenerierten Cellulose Langperioden vorliegen, wenn sie auch meist erst durch eine Vorbehandlung röntgenographisch sichtbar werden.

Alle Meridianreflexe, sowohl von den synthetischen Hochpolymeren als auch von Cellulose, zeigen auffällige Eigenschaften, die sie von den aus dem Weitwinkeldiagramm bekannten Reflexen unterschieden. Zuerst fällt auf, daß sie sehr diffus sind und daß dementsprechend höhere Ordnungen meist fehlen. Reflexe höherer Ordnung wurden bisher nur ganz vereinzelt beobachtet, und zwar von HESS und KIESSIG (47) bei Polyurethan und Polyamiden, sowie von STATTON (57), SELLA (58) und MANDELKERN (59 u. 59a) an Polyäthylen. Sie treten aber nur in geringer Intensität auf und stellen jedenfalls eine Ausnahme von der Regel dar.

Ferner steht fest, daß die Langperiode keinen Zusammenhang mit der chemischen Strukturperiode aufweist. Es kann sich daher nicht um ein kristallographisches Übergitter handeln. Das geht auch daraus hervor, daß die Orientierung der Kleinwinkelmeridianreflexe durchaus nicht immer mit der des Weitwinkeldiagramms zusammenfällt. Wenn Abweichungen auftreten, ist in der Regel das Kleinwinkeldiagramm besser orientiert. Die Langperiode ist auch vom Molekulargewicht unabhängig, kann also nicht direkt mit den durch die Molekülenden bewirkten Fehlstellen in Verbindung gebracht werden.

Die Meridianreflexe erscheinen normalerweise quer zum Meridian strichförmig auseinandergezogen. Es handelt sich also um Schichtlinienreflexe, die nicht mit den durch mangelhafte Orientierung bedingten Sicheln verwechselt werden können. Durch die Arbeiten von ARNETT, MEIBOHM und SMITH (48) wurden auch sog. Vierpunktdiagramme bekannt, die sich seither ebenfalls als eine recht allgemeine Erscheinung herausgestellt haben. Es handelt sich dabei um Reflexe, deren Intensitätsverteilung auf der Schichtlinie rechts und links vom Meridian je ein Maximum besitzt, so daß im Diagramm der Eindruck von vier Flecken entsteht (oben und unten je zwei). Das bekannteste Beispiel liefert das Polyäthylen. Sowohl Schichtlinien- als auch Vierpunktdiagramme weisen auf ein lineares Gitter mit nur geringen seitlichen Dimensionen der Fibrillen hin. Der Zusammenhang der Schichtlinien mit dem Querschnitt der Fibrillen konnte von HESS und KIESSIG (47) augenfällig demonstriert werden. An gewalztem Perlon, in dem dadurch eine höhere Orientierung hergestellt ist, ist die Länge der Schichtlinien deutlich verschieden, je nachdem man das Präparat normal oder parallel zur Walzebene durchstrahlt. Darin drückt sich der anisometrische Querschnitt der Fibrillen aus (vgl. I3c). Eine ähnliche Untersuchung wurde von STATTON (60) durchgeführt, auf die später noch in anderem Zusammenhang eingegangen werden wird.

Große Schwierigkeiten bereitet dem Verständnis die Abhängigkeit der großen Perioden von der Vorbehandlung, Tempern, Dehnen und Quellen. Es ändert sich dabei nicht nur die Lage des Reflexes, sondern auch die Schärfe, Intensität und die Länge der Schichtlinie sowie bei Vierpunktdiagrammen der Winkel, um den das Intensitätsmaximum vom Meridian absteht, und zwar meist irreversibel. Auch kann ein Schichtliniendiagramm sich in ein Vierpunktdiagramm umwandeln. Solche Veränderungen sind von allen auf dem Gebiet tätigen Forschern beobachtet worden. Die Erscheinungen sind aber zu vielfältig und unübersichtlich, um eine einfache Darstellung zu gestatten. Im allgemeinen läßt sich sagen, daß eine Erhöhung der Temperungstemperatur die Periode vergrößert. Gleichzeitig steigt meist auch die Intensität und die Schärfe des Reflexes, während er sich auf der Schichtlinie zusammenzieht. Es läßt sich schwer sagen, ob die Intensitätserhöhung für sich besteht oder ob sie nur eine indirekte Folge des Zusammenziehens des Reflexes auf ein kleineres Gebiet ist.

Die Veränderung der Langperiode durch Hitzebehandlung ist irreversibel. Reversible Änderungen werden dagegen durch Quellen und Dehnen, letzteres allerdings teilweise auch irreversibel, hervorgerufen. So berichtet ZAHN (50) von Perlon U (Polyurethan) eine reversible Vergrößerung der Periode durch Quellen mit 5%iger Phenollösung bei gleichzeitiger Verkürzung der Schichtlinie. Ein ähnlicher Effekt tritt durch Dehnen ein, und zwar genügen bereits 10% Dehnung, um die Periode von etwa 80 Å auf etwa 100 Å aufzuweiten.

In diesem Zusammenhang sind Untersuchungen besonders wünschenswert, die das Verhalten einer Substanz über den ganzen Dehnungsbereich systematisch verfolgen. Hierzu liegen Arbeiten von HENDUS (61) sowie von GUINIER und BELBÉOCH (24) vor. Beide befassen sich mit Polyäthylen, die erstere nur mit verzweigtem Hochdruckpolyäthylen, die letztere mit einer Reihe von verschiedenen Präparaten. Die Ergebnisse stimmen im wesentlichen überein. Der diffuse Ring des unorientierten Präparates zieht sich beim Dehnen in Form von Sicheln am Meridian zusammen. Erst nach der Striktion (necking) tritt das Schichtliniendiagramm auf, das schließlich in das bekannte Vierpunktdiagramm übergeht. GUINIER und BELBÉOCH finden dabei eine deutliche Abhängigkeit der Periode von der Temperatur, bei der die Dehnung vorgenommen wird. Sie beträgt bei 58° C 125 Å, bei 66° C 135 Å und bei 95° C 170 Å.

Nach der Relaxation bei 50° C liegt ein Vierpunktdiagramm mit einer Periode von 140 Å vor. Interessant ist, daß das ganze Verhalten des Kleinwinkeldiagramms nur wenig von der Qualität der Präparate abhängt (melt index), nur Niederdruckpolyäthylen macht eine Ausnahme, indem es im Ausgangszustand keinen diffusen Ring zeigt. Die beim Dehnen beobachteten Änderungen des Weitwinkeldiagramms stehen in keinem klaren Zusammenhang mit denen der KWS. Insbesondere zeigt sich kein unstetiger Übergang bei der Striktion. Andrerseits scheint aber unzweifelhaft, daß das Kleinwinkeldiagramm wesentlich vom Vorhandensein kristalliner Bereiche bedingt ist, denn im völlig amorphen Zustand tritt auch keine KWS auf.

b) Schichtkristalle

Ein ganz neuer Gesichtspunkt in der Frage der morphologischen Struktur der festen Hochpolymeren ist neuerdings durch die Entdeckung der Schichtkristalle aufgetaucht. 1957 stellten unabhängig voneinander FISCHER (62), KELLER (63), SELLA (58) und TILL (64) fest, daß lineares Polyäthylen (Marlex 50) aus einer Lösung in Toluol in Form von flachen Tafeln kristallisiert, die ein deutliches Stufenwachstum zeigen. Häufig sind auch Schraubenversetzungen zu beobachten. KELLER konnte durch Elektronenbeugung nachweisen, daß die Ketten normal zur Tafelebene stehen. Nachdem die Stufenhöhe aber nur in der Größenordnung von 100 Å liegt, gelangte KELLER zu dem zwingenden Schluß, daß die Fadenmoleküle gefaltet sein müssen. Zu demselben Ergebnis war übrigens bereits STORKS (65) durch seine Untersuchungen an Guttapercha gekommen. Stufenkristalle wurden neuerdings auch bei isotaktischem Poly-4-methyl-penten-1 (66) sowie bei Polyurethan und verschiedenen Polyamiden (67) festgestellt. Besonders die letzteren Arbeiten aus dem Kreis von STUART machen es wahrscheinlich, daß Kristallamellen bereits beim Kristallisieren aus der Schmelze auftreten und eine wichtige Rolle bei der Bildung der Sphärolithe spielen.

Das Wachstum in Form von Kristallamellen muß auch im Kleinwinkeldiagramm zum Ausdruck kommen. Tatsächlich konnten EPPE (67) und KELLER (68) an Lamellenbündeln von Marlex 50, die aus heißer Xylollösung gewonnen wurden, vier Reflexe in Form von Ringen erhalten, die als die vier ersten Ordnungen zu einer Periode von 120 Å gehören. An aufgewachsenen orientierten Lamellen wies KELLER nach, daß sich die Reflexe normal zur Lamellebene anordnen, also der Schichtdicke zuzuschreiben sind. Ähnliche Ergebnisse wurden von SELLA (58) auch an aus der Schmelze erstarrten Proben erhalten. Ein Unterschied besteht aber insofern, als nur in den Bereichen an der Oberfläche dieselbe Periode wie in den Einzelkristallen auftritt. Im Innern des Präparates hingegen findet SELLA größere Werte bis zu 400 Å. Ebenso

ergeben sich Abhängigkeiten vom Verzweigungsgrad und von der Nachbehandlung. Ein von Eppe (67) aus der Schmelze auf —70° C ab-
geschrecktes Polyurethan zeigte zunächst keine KWS. Erst nach Tempern bei 80° bzw. 170° ergaben sich Reflexe entsprechend 50 bzw. 150 Å,
deren Intensität mit der Behandlungszeit stieg, ohne daß sich die Lage
weiter änderte.

Daß auch bei lamellaren Einkristallen die Schichtdicke sich mit der
Temperatur ändert, bei der die Ausscheidung aus der Lösung erfolgt,
geht aus einer weiteren Arbeit von Keller und O'Connor (69) hervor.
Im Temperaturintervall von 10 bis 90° C verschiebt sich der Wert für
die Periode von 92 bis 140 Å. Die Verhältnisse liegen hier also ähnlich wie
bei den Langperioden der gezogenen Fasern.

Die Einkristalle von Polyäthylen entsprechen weitgehend denen der
Paraffine. Nachdem bis zu $C_{100}H_{202}$ die Moleküle im Kristall gestreckt
bleiben, ist die Frage von Interesse, von welcher Moleküllänge an die
Faltung eintritt. Da eine Synthese von noch höheren Paraffinen schwierig
ist, versuchten Keller und O'Connor (69) dem Problem dadurch näher
zu kommen, daß sie Polyäthylen dem thermischen Abbau unterwarfen.
Wie erwartet, erwies sich die Dicke der Kristallstufen, also die Länge
der Falten, als unabhängig vom Abbaugrad, solange die Moleküle nur
noch genügend lang waren. Bei noch stärker abgebauten Produkten aber,
die sich bereits durch ihre wachsige Beschaffenheit als Paraffine zu erkennen gaben, zeigten sich nach Fraktionierung kleinere Perioden bis 40 Å
herunter. Man wird annehmen dürfen, daß diese Werte der Kettenlänge
entsprechen, wenn auch in dieser Arbeit ein strenger Beweis nicht geführt
wurde. Immerhin spricht die Kontrolle durch den Schmelzpunkt sehr dafür.

Der umgekehrte Weg, nämlich der Aufbau von immer längeren
Molekülen, ist in einer ganz neuen Arbeit von Zahn (70) eingeschlagen.
In der Reihe der Carbobenzoxy-Polymainocapronsäuren ergab sich, daß
die Langperiode nach anfänglicher Zunahme von $n = 6$ bis $n = 12$ mit
Werten von 67 bis 74 Å nahezu konstant bleibt, während die Molekül-
längen von 60 Å bei $n = 6$ auf 111 Å bei $n = 12$ steigen. Dieses Ergebnis
entspricht durchaus den Verhältnissen beim Polyäthylen. Auffällig erscheint jedoch, daß bei dem Oligomeren mit $n = 6$ die Moleküllänge von
60 Å noch kleiner ist als die Periode von 67 Å, was anschaulich kaum
zu verstehen ist. Doch ist die Diskrepanz auch wieder nicht so groß, als
daß sich nicht vielleicht eine natürliche Erklärung finden ließe. Man wird
etwa daran denken können, daß die Berechnung der Periode nach dem
Braggschen Gesetz bei Hochpolymeren nicht ganz streng ist (Störungen
II Art). Jedenfalls sind hier weitere Untersuchungen abzuwarten.

In den Reihen vom Nylon- und Terylen-Typ traten keine Klein-
winkel-Meridianreflexe auf, obwohl genügend lange Moleküle erreicht
wurden. Auch hier ist eine Erklärung noch offen.

c) Deutung der Meridianreflexe

HESS und KIESSIG (46) haben zur Erklärung der Langperioden ihr bekanntes Modell entwickelt, das heute weitgehend angenommen ist und wohl die plausibelste Deutung darstellt. Es besteht aus einem fibrillären Schema mit durchlaufenden Fadenmolekülen, in dem besser und schlechter geordnete Bereiche in mehr oder minder guter Regelmäßigkeit abwechseln. Dem üblichen Sprachgebrauch folgend, kann man von kristallinen und amorphen Bereichen sprechen, ohne sich bezüglich der Problematik dieser Begriffe bei den Hochpolymeren festzulegen. Die Langperiode würde der Summe der Längen von je einem kristallinen und einem amorphen Bereich entsprechen. Natürlich muß man auch eine ziemlich hohe statistische Schwankung annehmen, um die diffuse Ausbildung der Reflexe und das Fehlen höherer Ordnungen zu erklären.

Die stärkste Stütze für das Modell stellen ohne Zweifel die elektronenoptischen Befunde von HESS und MAHL (71) dar, durch die bei Polyvinylalkohol und bei Cellulose die Periodizität durch Behandlung mit Jod und Thallium direkt sichtbar gemacht werden konnte. Eine gewisse Diskrepanz besteht nur noch in den Zahlenwerten. Neben der gewöhnlichen aus der KWS bekannten Periode treten noch größere bis zu 700 Å auf, und zwar in noch stärkerer Ausprägung. Im Kleinwinkelgebiet sind solche Superperioden noch nicht beobachtet worden. Doch mag dies daran liegen, daß die entsprechenden Reflexe bei extrem kleinen Winkeln auftreten müssen und daher experimentell schwer festzustellen sind. Es bleibt aber die Frage, warum die „normale" Langperiode in der Gegend von 165 Å im Elektronenmikroskop die Ausnahme und nicht die Regel darstellt. Abgesehen von diesen Unstimmigkeiten, ist jedenfalls die grundsätzliche Möglichkeit einer großen Periodizität in festen Hochpolymeren überzeugend dargetan.

Nun ist aber die Deutung der Meridianreflexe selbst durch das Modell von HESS und KIESSIG noch nicht restlos befriedigend gelungen. Vor allem die starke Abhängigkeit der Reflexe von der Nachbehandlung bereitet große Schwierigkeiten, wie die Autoren selbst betont haben. Eine Veränderung der Periode durch Wachsen der kristallinen auf Kosten der amorphen Bereiche ist nur innerhalb sehr enger Grenzen möglich und kann die tatsächlich beobachteten Verschiebungen nicht erklären. Man ist gezwungen, mit HESS und KIESSIG einen totalen Umbau der Textur anzunehmen. Auch GUINIER und BELBÉOCH (29) kommen bei ihren bereits erwähnten Untersuchungen an Polyäthylen zu dem Ergebnis, daß man dieser Konsequenz nicht ausweichen kann. Das mag für die irreversiblen Änderungen zutreffen, ist aber bei den reversiblen Änderungen kaum denkbar. Es sei hier etwa an die Befunde von ZAHN (50) erinnert, wo bei Perlon U durch eine Dehnung von bloß 10% eine Aufweitung der Periode um 23% eintritt. Eine Erklärung durch ein

Auseinanderrücken der Kristallite oder durch eine reversible Verlän-
gerung der kristallinen Bereiche auf Kosten der amorphen kommt hier
nicht in Frage. Es bleibt also noch ein offenes Problem.

Eine weitere, wenn auch nicht unüberwindliche Schwierigkeit bilden
die Vierpunktdiagramme. Sie weisen darauf hin, daß Netzebenen vorliegen, die gegen die Faserachse geneigt sind. Die Annahme einer schrägen
Lage der Grenzfläche zwischen kristallinen und amorphen Bereichen
innerhalb einer Fibrille reicht zur Erklärung nicht aus, da wir die Quer-

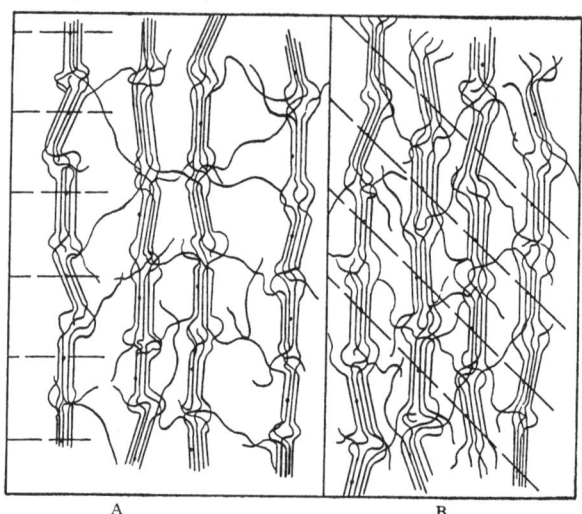

A   B

Abb. 4A u. B. Schema des Modells von HESS und KIESSIG in der Modifikation von STATTON. A. entspricht
Schichtlinienreflex. B. entspricht Vierpunktdiagramm. [Nach W. O. STATTON und G. GODARD (60)]

dimension jedenfalls als klein betrachten müssen. STATTON (60, 72) hat
diese Frage eingehend diskutiert. Er gelangt zu einer leichten Modifikation des Modells, indem er den Elementarfibrillen mehr Selbständigkeit
zuordnet und sie als gegeneinander verschiebbar annimmt. Wie Abb. 4
zeigt, kann man dann leicht verstehen, daß sich schiefgestellte Netzebenen ausbilden, wenn sich die kristallinen Bereiche benachbarter
Fibrillen nicht in gleicher Höhe befinden. Ja man sollte dies sogar als
den Normalfall ansehen, weil aus sterischen Gründen eine Anordnung in
Ebenen normal zur Faserachse nicht zu erwarten ist. Diese kommt auch
deswegen nicht in Frage, weil sie mit dem normalen Schichtliniendiagramm im Widerspruch steht, das ja geringe seitliche Dimensionen
der Netzebenen fordert. In diesem Fall bleibt nur die Annahme einer
statistisch regellosen Position der Elementarfibrillen.

Mit den Betrachtungen von STATTON ist die Frage der lateralen
Ordnung angeschnitten. Die allgemeinste und umfassendste Behandlung
liegt in der Theorie des Parakristalls von HOSEMANN (41) vor (s. 13c),

die auch auf die Kleinwinkelreflexe anwendbar ist (73). Ein hierfür besonders geeignetes Analysenverfahren ist von BONART (74) angegeben worden. Nachdem über den ganzen Fragenkomplex an dieser Stelle ein umfassender Bericht von BONART erscheinen soll, kann auf eine nähere Besprechung verzichtet werden. Es sei nur festgehalten, daß die laterale Ordnung auf alle Fälle einen wesentlichen Gesichtspunkt bei der Erklärung der Meridianreflexe darstellt. Qualitativ ist dies durch die lichtoptischen Modellversuche in den zitierten Arbeiten erwiesen.

Die Entdeckung der lamellaren Einkristalle und der Kettenfaltung legt den Gedanken nahe, daß diese in einem näheren Zusammenhang mit den Langperioden der gezogenen Fasern stehen könnte, zumal die Werte für die Periode und für die Faltenlänge gut übereinstimmen. KELLER (75) hat auf Grund seiner Versuche an gedehntem Polyäthylen den Vorschlag gemacht, daß die Periodizität der Kettenfaltung auch in den Elementarfibrillen bestehen bleibt, sei es, daß beim Ausziehen der Moleküle stufenartige Knickstellen zurückbleiben oder daß die Falten teilweise überhaupt nicht aufgelöst werden. Nach dieser Vorstellung würden die amorphen Bereiche des Modells von HESS und KIESSIG entweder überhaupt wegfallen oder aber durch die Stufen oder Falten in den Einzelmolekülen bedingt sein. Bisher existiert weder ein überzeugender Beweis für noch gegen die Annahme von KELLER. Auch ist der ganze Mechanismus des Ausziehens der Fibrillen noch unklar.

Schon mehrfach (46, 59, 76) ist der Gedanke aufgetaucht, daß die Periodizität in der morphologischen Struktur mit den Wärmeschwingungen in Zusammenhang stehen könnte. Nun haben neuerdings FISCHER und PETERLIN (77) eine Theorie entwickelt, die diesen Gedanken quantitativ zu formulieren sucht. Der Gedankengang ist, kurz skizziert, folgender: Der Zusammenhalt der monomeren Reste innerhalb eines Fadenmoleküls ist sicher viel stärker als der zwischen benachbarten Ketten. Man kann daher annehmen, daß die Längs- und Rotationsschwingungen benachbarter Moleküle weitgehend voneinander unabhängig sind. Es folgt dann, daß die mittlere quadratische Abweichung eines Kettengliedes von der Ruhelage mit zunehmender Länge des Kristalls in Kettenrichtung immer größer wird. Dadurch ergibt sich eine zunehmende Verschmierung des Gitterpotentials, die zu einem Anstieg der freien Energiedichte des Kristalls führt. Andererseits nimmt die Energie pro Volumeneinheit infolge des Beitrages der Grenzflächenenergie mit wachsender Kristallitgröße ab. Diese beiden gegenläufigen Abhängigkeiten führen im allgemeinen zu einem Minimum der freien Energiedichte bei einer bestimmten, von den zwischenmolekularen Kräften und der Temperatur abhängigen Kristallitlänge. Nun würde eine exakte Berechnung zwar eine genaue Kenntnis der Schwingungen und zwischenmolekularen Kräfte erfordern, wie sie natürlich nicht zur Verfügung

steht. FISCHER und PETERLIN konnten aber unter plausiblen ver-
einfachenden Annahmen zeigen, daß eine Kristallitlänge in der Größenordnung von 100 Å folgt, in guter Übereinstimmung mit den Befunden
sowohl an den lamellaren Einkristallen als auch an den gezogenen Fasern.

d) Zusammenhang der Meridianreflexe mit dem Weitwinkeldiagramm

Die von dem Modell von HESS und KIESSIG postulierte sehr kleine
Kristallitlänge kann auch durch das Weitwinkeldiagramm überprüft
werden. Wie WALLNER (78) an Perlon L feststellte, gehorchen die Basisreflexe nicht streng dem Braggschen Gesetz. Die aus den verschiedenen
Ordnungen berechneten Werte für die chemische Strukturperiode zeigen
deutliche Unterschiede (16,25—17,2 Å; wahre Faserperiode 17,08 Å).
WALLNER erklärte diesen zunächst überraschenden Befund damit, daß
bei sehr kleinen Kristalliten die Lage eines Reflexes nicht nur wie ge-
wöhnlich vom Gitterfaktor allein, sondern auch vom Strukturfaktor der
Basiszelle abhängt. Ein deutlicher Effekt ist dann zu erwarten, wenn
letzterer selbst mit dem Winkel schnell veränderlich ist, während der
Gitterfaktor nur verhältnismäßig breite Maxima aufweist. Nachdem die
Atompositionen im Polyamidgitter bekannt sind, konnte WALLNER die
Rechnung in Strenge durchführen und zeigen, daß die Länge der
Ordnungsbereiche nur 4—5 Gitterzellen entsprechend 68,3—85,4 Å betragen kann. Das steht in bestem Einklang mit dem Kleinwinkel-
meridianreflex, aus dem eine Langperiode von 88 Å folgt. Einer allgemeinen Anwendung der Methode von WALLNER steht leider der Umstand entgegen, daß sie eine sehr genaue Kenntnis der Atompositionen
erfordert und daß deutliche Effekte nur in speziellen Fällen zu erwarten
sind.

Ein anderer Weg zur Kontrolle der Kristallitlängen besteht in der
Linienbreitenmessung der Basisreflexe. Bei der Cellulose liegt eine ältere
Messung von HENGSTENBERG und MARK (79) vor, nach der die Kristallitlänge von nativer Cellulose (Ramie) mindestens 600 Å und von regenerierter Cellulose mindestens 300 Å betragen sollte. Diese Angaben sind
allerdings unverträglich mit den aus der KWS bestimmten Langperioden
in der Gegend von 170 Å, passen aber recht gut auf die von HESS und
MAHL im Elektronenmikroskop gefundenen Riesenperioden. Vielleicht
wird man dieser Diskrepanz durch die Annahme ausweichen können, daß
die von HESS und KIESSIG vorgeschlagene Modellvorstellung nur auf
einen Teil der Elementarfibrillen zutrifft. Dieser Eindruck wird ja auch
durch die elektronenmikroskopischen Untersuchungen hervorgerufen.

Besonders eingehende Untersuchungen haben STATTON und GODARD
(60) an Terylen durchgeführt. Es wurden hochkristalline Filme mit
höherer Orientierung hergestellt und in Streifen übereinander gelegt.
Diese Pakete gaben bei Durchstrahlung in den drei Hauptrichtungen

(Zugrichtung = "end", normal dazu und parallel der Filmebene = "edge", normal zur Filmebene = "through") ganz verschiedene Weitwinkel- und Kleinwinkeldiagramme, die in Abb. 5 modellmäßig zusammengestellt sind. Für die Frage der Langperioden sind nur die Durchstrahlungen

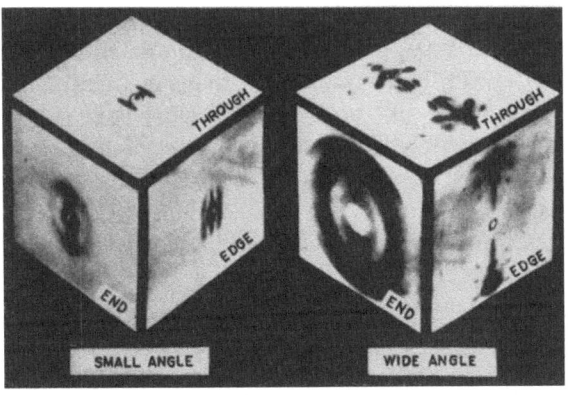

Abb. 5. Kleinwinkel- und Weitwinkeldiagramm von höher orientiertem Terylen. Die drei Hauptdurchstrahlungsrichtungen sind im Modell zusammengestellt. [Nach W. O. STATTON und G. GODARD (60)]

normal zur Zugrichtung maßgebend. Bei "edge" ergab sich ein Vierpunktdiagramm, bei "through" ein Schichtlinienreflex, beide in der Langperiode von 121 bis 124 Å übereinstimmend. In der Modellvorstellung von STATTON besagt dieses Ergebnis, daß die Elementarfibrillen nur in der Packung normal zur Filmebene eine regelmäßige Versetzung zeigen.

Aus der Linienbreite der entsprechenden Weitwinkelreflexe folgen die Abmessungen der kristallinen Bereiche: 74 Å in der Faserrichtung, 44 × 62 Å im Querschnitt. Es fällt dabei auf, daß der Querschnitt gar nicht so an-

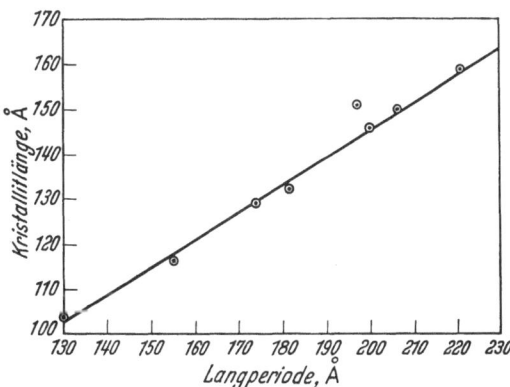

Abb. 6. Zusammenhang zwischen Kristallitlänge und Langperiode bei Polyäthylen. [Nach W. O. STATTON (52)]

isometrisch ist, wie man vielleicht nach der leichten Herstellung der höheren Orientierung erwarten könnte. Die Werte sind aber auch durch die gleichzeitige Vermessung der Schichtlinienlänge gestützt. Die Differenz zwischen der Langperiode (124 Å) und der Kristallitlänge (74 Å), also 50 Å, sollte der Länge der amorphen Bereiche entsprechen.

26a

Eine solche Zahlenangabe setzt natürlich einen scharfen Wechsel der Ordnungszustände voraus. Aber auch, wenn man einen solchen nicht annehmen will, stellt sie jedenfalls eine Kennzahl der Textur dar.

Die folgenden Versuchsserien von Statton (52, 72) an verschiedenen Proben von Terylen, Polyäthylen und Nylon-66 ergaben im wesentlichen das gleiche Bild. Durchaus ergeben sich aus der Linienbreite kleinere Werte der Kristallitlänge, als die Langperiode nach dem Kleinwinkel-meridianreflex beträgt, und zwar besteht zwischen beiden Größen innerhalb einer Serie eine recht gute lineare Abhängigkeit, wie Abb. 6 für Polyäthylen zeigt. Diese Untersuchungen stellen die bisher stärkste röntgenographische Stütze für das Modell von Hess und Kiessig dar.

3. Orientierte Systeme, äquatoriale Kleinwinkelstreuung

a) Äquatoriale Streukurven

Eine Strukturvorstellung, wie sie sich im Modell von Hess und Kies-sig, besonders in der von Statton spezifizierten Form ausdrückt, scheint zwangsläufig auch in seitlicher Richtung eine gewisse Regelmäßigkeit zu fordern. Die bisher vorliegenden Untersuchungen beziehen sich fast ausschließlich auf die Cellulose. Im Zusammenhang mit dieser Substanz hat vor allem Kratky (9) auf die Notwendigkeit einer Ordnung in kleinen Bereichen hingewiesen und zum erstenmal den Begriff des dichtgepackten Systems in die Kleinwinkelforschung eingeführt.

Die erste Frage ist, welche Evidenz für diese Anschauung vorliegt. Die Existenz von individuellen Bereichen mit anisometrischem Querschnitt (Micellen oder Fibrillen) konnte durch die Tatsache erwiesen werden, daß es Burgeni und Kratky (29) gelang, durch Walzen eine höhere Orientierung herzustellen, die auch im Kleinwinkeldiagramm direkt sichtbar gemacht werden konnte. Daß es sich aber nicht um eine Partikelstreuung handelt, geht daraus hervor, daß die Streukurve bei der Quellung sich ändert, wie von Kratky, Sekora und Treer (80) gezeigt und auch von Heyn (80b) bestätigt wurde. Auch in der bereits (II 1) besprochenen Arbeit von Kratky u. Mitarb. (42) war der Unterschied der dichten und gequollenen Fäden ganz deutlich. Es scheint somit sicher, daß die Streukurve von ungequollenen Fäden wesentlich durch die interpartikulären Interferenzen bestimmt ist. Auch die andere Feststellung von Heyn (81), daß schon die Streukurven von mäßig gequollenen Fäden der Guinierschen Näherung folgen sollen, bildet kein Gegenargument, da sie nicht bestätigt werden konnte und auf unbekannten speziellen Bedingungen zu beruhen scheint.

Als der überzeugendste Beweis für eine laterale Ordnung in der Cellulose und damit auch für die Existenz von ziemlich einheitlichen Elementarteilchen müssen die Maxima angesehen werden, die von Her-

MANS u. Mitarb. (82) bei vielen Kunstseiden am Äquator gefunden wurden. Wie Abb. 7 zeigt, variiert die Ausbildung von der Andeutung einer Inflexion bis zu einem recht scharfen Reflex, ohne daß ein Zusammen-

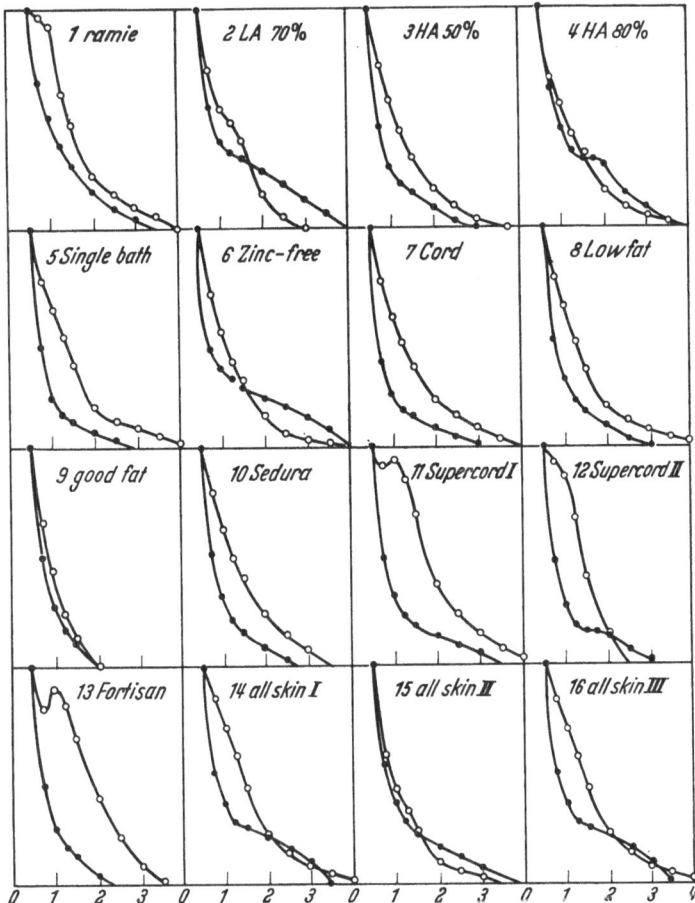

Abb. 7. Aquatoriale Kleinwinkelstreuung von verschiedenen Cellulosefasern. ● trocken; ○ wassergequollen.
[Nach D. HEIKENS, P. H. HERMANS, P. F. VAN VELDEN und A. WEIDINGER (82)]

hang mit den textilen Eigenschaften der Fasern ersichtlich wäre. Außerdem fällt auf, daß stets ein diffuser Untergrund in Form einer monoton fallenden Streukurve vorhanden ist.

Die Maxima erscheinen normalerweise erst nach leichtem Aufquellen mit Wasser oder LiOH-Lösung. Nachdem es sich durchwegs um hoch orientierte Präparate handelt, ist bei den Aufnahmen mit Spaltblende der Lorentzfaktor der Länge bereits eliminiert, so daß man auf das Maximum unmittelbar das Braggsche Gesetz anwenden kann (natürlich nur näherungsweise, da bestimmt keine ideale Periodizität vorliegt). Es ergeben sich so Perioden von 70 bis 90 Å je nach Faser und Behandlung. Vereinzelt wurden auch bei trockenen Fasern allerdings sehr intensitätsschwache Äquatorreflexe festgestellt, die durchwegs kleineren Perioden von 50 bis 60 Å entsprechen. Nachdem die Periode die Summe von Teilchendicke und Zwischenraum darstellt, ist der Unterschied begreiflich. Wir können die Werte der trockenen Fasern als Maß für die Querdimensionen der Elementarfibrillen ansehen.

Die Beobachtungen von Hermans sind von zahlreichen Forschern bestätigt und erweitert worden, so von Heyn (80, 83), Fournet (84), Kratky (85), Kiessig (55) und Statton (56). Es handelt sich also um eine gesicherte Eigenschaft der regenerierten Cellulose. Native Cellulose verhält sich merklich anders. Soweit überhaupt Äquatorreflexe gefunden wurden (84, 85), ergeben sie wesentlich größere Perioden, so Ramie 110 Å und Baumwolle 126 Å. Auch scheinen die Maxima viel diffuser zu sein. Nimmt man noch dazu, daß bisher auch keine Meridianreflexe gefunden worden sind, so folgt doch ein merklicher Unterschied in der morphologischen Struktur von nativer und regenerierter Cellulose.

Was nun das Hervorrufen der Äquatorreflexe durch Quellungsbehandlung betrifft, so werden wir uns wohl der Ansicht von Hermans anschließen müssen, daß die laterale Ordnung auch in den trockenen Fasern bereits vorhanden ist, wenn sie auch im Kleinwinkeldiagramm nicht in Erscheinung tritt. Das kann daran liegen, daß die Intensität infolge zu kleiner Zwischenräume nicht ausreicht oder daß die Elementarfibrillen seitlich verklebt sind. Jedenfalls ist kein grundlegender Umbau der Textur anzunehmen. Das geht aus der Reversibilität der Erscheinungen hervor. Hermans konnte so z. B. bei einer säurebehandelten Fortisanfaser (hoch orientierte, verseifte Celluloseacetatfaser) die diffuse Streuung des trockenen und das ausgeprägte Maximum des gequollenen Präparates beliebig oft reversibel durch Quellen und Trocknen in einander umwandeln.

b) Hohlräume; Auswertung der Streukraft

In der Regel zeigt sich am Äquator nur eine diffuse Streuung und diese bildet auch dann den Hauptanteil, wenn zusätzlich Maxima vorhanden sind. Nachdem eine vollständige Auswertung auf die bei dichtgepackten Systemen bekannten Schwierigkeiten stößt, hat es sich als zweckmäßig erwiesen, auf eine modellmäßige Interpretation zu verzichten und sich nur auf einen allgemeinen Parameter zu beschränken. Als solcher hat im Zusammenhang mit der Cellulose vor allem die Streukraft (13a) eine gewisse Bedeutung gewonnen. Sie sagt zwar nichts über die geometrische Struktur, wohl aber etwas über die materielle Zusammensetzung des Systems aus.

Als erster Versuch in dieser Richtung — allerdings ohne den Begriff der Streukraft zu benützen — können die Arbeiten von KRATKY u. Mitarb. (*80, 86*) betrachtet werden. Sie konnten zeigen, daß mit dem Austausch des Quellungsmittels sich die Intensität der KWS der Cellulose sehr stark ändert, und zwar proportional mit dem Quadrat der Elektronendichtendifferenz. Dieser Befund kann nur so gedeutet werden, daß das Quellungs-

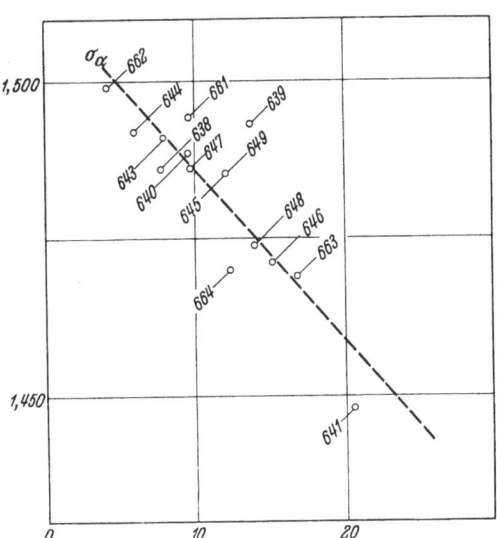

mittel bereits vorhandene Hohlräume ausfüllt, deren Existenz daher damit bewiesen wird. Daß die Hohlräume vorhanden und gerade für die diffuse Streuung verantwortlich sind, ist auch sonst wahrscheinlich. Besonders STATTON (*56, 57*) vertritt nachdrücklich diesen Standpunkt.

Als erster hat STERN (*87*) die Streukraft selbst zur Klärung dieser Frage bei der Cellulose herangezogen. Seine Untersuchungen sind durch HERMANS u. Mitarb. (*88*) weitergeführt und wesentlich erweitert worden. Vor allem haben sie auch

Abb. 8. Zusammenhang zwischen Streukraft und Dichte verschiedener Cellulosefasern. [Nach P. H. HERMANS, D. HEIKENS und A. WEIDINGER (*88*)]

den Absolutwert der Streukraft bestimmt, während von Stern nur Relativwerte angegeben werden. Im übrigen sind die Ergebnisse aber miteinander verträglich.

Das erste wichtige Resultat der Untersuchung von HERMANS ist, daß die Streukraft der meisten Cellulosefasern wesentlich höher ist, als einem reinen Zweiphasensystem bestehend aus kristallinen und amorphen Anteilen entsprechen würde. Wie Abb. 8 zeigt, besteht eine deutliche Antibasie zwischen Streukraft und spezifischem Gewicht, wenn auch ein strenger funktioneller Zusammenhang nicht angegeben werden kann. Durch Vergleich des dichtesten Präparates 662, das als einziges normale Streukraft zeigt, mit dem am wenigsten dichten 641 berechnet HERMANS für letzteres einen Volumanteil der Hohlräume von 3,5% aus der Dichte. Aus der Streukraft ergibt sich aber nur 0,78%. Die Diskrepanz erklärt sich zwanglos durch die Annahme, daß nur ein Bruchteil der Hohlräume in kolloiden Dimensionen liegt und röntgenographisch in Erscheinung tritt, während bei der Dichte natürlich auch die großen berücksichtigt sind. Damit wird auch die beträchtliche Schwankung der Meßpunkte in Abb. 8 verständlich. Sie bedeutet einfach, daß bei verschiedenen Cellulosefasern die Größenverteilung der Mikrohohlräume stark variiert.

Bei der Veränderung der Streukraft mit der Feuchtigkeitsaufnahme sind die Verhältnisse nicht so übersichtlich. Man würde erwarten, daß nach dem vollständigen Auffüllen der Hohlräume die Streukraft wieder auf den Wert für ein Zweiphasensystem herabsinken sollte. Dieses Verhalten zeigt nur ein Teil der von HERMANS untersuchten Proben. Meist liegt die Streukraft viel höher. HERMANS schlägt die plausible Deutung vor, daß neue Hohlräume gebildet werden. Doch bleibt hier eine weitere Bestätigung abzuwarten.

Alle diese Untersuchungen beziehen sich auf die Cellulose. Es kann aber keine Frage sein, daß die Methode grundsätzlich auf alle Hochpolymeren anwendbar ist und ähnliche Untersuchungen von Interesse wären. Bisher liegt ein Befund nur am Polyäthylen von GUINIER und BELBÉOCH (24) vor. Sie finden, daß der Anteil an Mikrohohlräumen höchstens 0,1% betragen kann. Es sind also die Verhältnisse bei der Cellulose nicht ohne weiteres auf andere Hochpolymere zu übertragen.

Zusammenfassung

Die Röntgenkleinwinkelstreuung ist ein Beugungseffekt an Inhomogenitäten von kolloiden Dimensionen. Sowohl ihre Intensität als auch die Art ihrer Winkelabhängigkeit geben Auskunft über die morphologische Struktur des Systems. Bei der Auswertung ist zwischen verdünnten und dichtgepackten Systemen zu unterscheiden, je nachdem ob die interpartikuläre Interferenz vernachlässigt werden kann oder nicht.

In verdünnten Systemen corpuscularer Teilchen kann die Streukurve durch eine Gaußsche Glockenkurve approximiert werden. Es lassen sich daraus der Streumassenradius, ein Maß für die mittlere Größe sowie das Molekulargewicht bzw. das Volumen ermitteln. Bei stäbchen- oder blättchenförmigen Teilchen tritt zusätzlich ein Lorentzfaktor auf. Nach dessen Eliminierung führt die Auswertung auf den Querschnitt bzw. die Blättchendicke.

In stark polydispersen Systemen kann nach HOSEMANN durch ein einfaches graphisches Verfahren die statistische Größenschwankung bestimmt werden. Ein Kriterium für die Anwendung dieser Methode auf dichtere Systeme wird diskutiert.

Die Kleinwinkelstreuung von dichtgepackten Systemen kann entweder modellmäßig oder hinsichtlich allgemeiner Parameter interpretiert werden. Letztere können sein: Streukraft und Invariante, Korrelationslänge und Kohärenzlänge sowie die innere Oberfläche. Ihr Zusammenhang mit der Streukurve und die Auswertungsmethode werden besprochen.

In Systemen mit Faserorientierung bezieht sich die Streukurve bei der üblichen Aufnahmetechnik mit strichförmigem Primärstrahl nur auf die morphologische Struktur in der Äquatorebene. Es wird gezeigt, daß eine eindeutige Beziehung zur Kleinwinkelstreuung am entsprechenden ungeordneten System besteht. Bei höherer Orientierung kann auch die Anisometrie des Querschnitts sichtbar gemacht werden, wenn man in Richtung der Faserachse durchstrahlt.

Das Auftreten von diskreten Maxima deutet auf eine regelmäßige Anordnung von gleichartigen Teilchen hin. Es ergibt sich die Notwendigkeit, Störungen zweiter Art in Betracht zu ziehen. Es wird gezeigt, daß sie sowohl beim linearen Gitter als auch beim Parakristall für die Breite der Maxima und das Fehlen höherer Ordnungen verantwortlich sind. Aus der Länge der Schichtlinienreflexe kann die seitliche Abmessung der beugenden Bereiche erschlossen werden.

Es wird eine Übersicht über die wichtigsten Ergebnisse der Kleinwinkelmethode an festen Hochpolymeren gegeben. Eine Auswertung als verdünntes System ist nur bei hochgequollenen Präparaten möglich und führt dann auf Teilchengrößen, so bei Cellulose, Seide und Alginaten.

Die diskreten Meridianreflexe werden hinsichtlich ihrer Ausbildung und Abhängigkeit von der Vorbehandlung besprochen. Sie stellen ein natürliches Beispiel für ein lineares Gitter mit Störungen zweiter Art dar. Dagegen muß bei den lamellaren Einkristallen eine regelmäßige Kettenfaltung angenommen werden. Auf den Zusammenhang mit den Basisreflexen der Oligomeren wird hingewiesen.

Die Deutung der Meridianreflexe nach dem Modell von HESS und KIESSIG, dem Parakristall von HOSEMANN und der thermodynamischen Theorie von FISCHER und PETERLIN wird diskutiert. Auch die vergleichenden Untersuchungen am Weitwinkel- und Kleinwinkeldiagramm bestätigen im wesentlichen die modellmäßigen Vorstellungen.

Die Arbeiten über die äquatoriale Streuung beziehen sich überwiegend nur auf die Cellulose. Es treten Maxima auf, die auf eine laterale Ordnung hinweisen, und ein diffuser Untergrund, der als Effekt der Mikrohohlräume anzusehen ist. Dieser Gesichtspunkt wird durch die quantitative Untersuchung der Streukraft durch HERMANS bestätigt.

Literatur

1. KIESSIG, H.: Kolloid-Z. **98**, 213 (1942).
2. — Kolloid-Z. **152**, 62 (1957).
3. HOSEMANN, R.: Z. Physik **113**, 751; **114**, 133 (1939).
4. KRATKY, O.: Z. Elektrochem. **58**, 49 (1954); **62**, 66 (1958). — KRATKY, O., u. Z. SKALA: Z. Elektrochem. **62**, 73 (1958).
5. GUINIER, A., et G. FOURNET: J. phys. radium **8**, 345 (1947). — DuMOND, J. M. W.: Phys. Rev. **72**, 83 (1947).
6. KRATKY, O., G. POROD u. L. KAHOVEC: Z. Elektrochem. **55**, 53 (1951).

7. GEROLD, V.: Acta Cryst. **10**, 287 (1957).

8. KRATKY, O., G. POROD u. Z. SKALA: Acta Phys. Austriaca **13**, 76 (1960).

9. — Naturwissenschaften **26**, 94 (1938); **30**, 542 (1942); Z. Elektrochem. **46**, 550 (1940).

10. HOSEMANN, R.: Z. Elektrochem. **46**, 535 (1940); Kolloid-Z. **117**, 13; **119**, 129 (1950).

11. KRATKY, O., u. G. POROD: Z. physik. Chem. **7**, 236 (1956).

12. GUINIER, A.: Compt. rend. **204**, 1115 (1937); Ann. phys. **12**, 161 (1939); J. chim. phys. **40**, 133 (1943).

13. POROD, G.: Kolloid-Z. **124**, 83 (1951). — KAHOVEC, L., G. POROD u. H. RUCK: Kolloid-Z. **133**, 16 (1953).

14. KRATKY, O., u. G. POROD: Acta Phys. Austriaca **2**, 133 (1948).

15. POROD, G.: Acta Phys. Austriaca **2**, 255 (1948); Z. Naturforsch. 4a, 401 (1949).

16. SHULL, C. G., and L. C. ROESS: J. Appl. Phys. **18**, 295 (1947). — MALMON, A. G.: Acta Cryst. **10**, 639 (1957). — SCHMIDT, P. W.: Acta Cryst. **8**, 772 (1955); **11**, 674 (1958). — BEIDL, G., M. BISCHOF, G. GLATZ, G. POROD, J. CH. v. SACKEN u. H. WAWRA: Z. Elektrochem. **61**, 1311 (1957).

17. GUINIER, A., and G. FOURNET: Small-Angle Scattering of X-Rays. New York: John Wiley & Sons, Inc. 1955.

18. ROESS, L. C., and C. G. SHULL: J. Appl. Phys. **18**, 308 (1947).

19. DEBYE, P.: Physik. Z. **28**, 135 (1927).

20. FOURNET, G.: Compt. rend. **228**, 1421, 1801 (1949); Acta Cryst. **4**, 293 (1951).

21. POROD, G.: Kolloid-Z. **125**, 51 (1952).

22. HOSEMANN, R.: Z. Elektrochem. **61**, 1008 (1957). — HOSEMANN, R., u. D. JOERCHEL: Kolloid-Z. **152**, 49 (1957). — HOSEMANN, R., u. R. BONART: Kolloid-Z. **152**, 53 (1957).

23. DEBYE, P., and A. M. BUECHE: J. Appl. Phys. **20**, 518 (1949).

24. BELBÉOCH, B., u. A. GUINIER: Makromol. Chem. **31**, 1 (1959).

25. DEBYE, P., H. R. ANDERSON jr. and H. BRUMBERGER: J. Appl. Phys. **28**, 679 (1957).

26. NORDSTRAND, R. A. VAN, and K. M. HACH: Amer. Chem. Soc. meeting, Chicago, Sept. 1953.

27. KRATKY, O., u. G. POROD: Z. Elektrochem. **58**, 918 (1954).

28. HERMANS, P. H., u. A. WEIDINGER: Z. Elektrochem. **58**, 924 (1954).

29. BURGENI, A., u. O. KRATKY: Z. physik. Chem. (B) **4**, 401 (1929).

30. KRATKY, O., u. S. KURIYAMA: Z. physik. Chem. (B) **11**, 363 (1931).

31. — u. A. SEKORA: J. makromol. Chem. **1**, 113 (1943).

32. BEAR, R. S.: J. Am. Chem. Soc. **64**, 727 (1942); **65**, 1784 (1943); **66**, 1297, 2043 (1944); **67**, 1625 (1945).

33. ZERNICKE, F., u. J. A. PRINS: Z. Physik **41**, 184 (1927).

34. KRATKY, O.: Physik. Z. **34**, 482 (1933); Monatsh. Chem. **76**, 311 (1947).

35. HERMANS, J. J.: Rec. trav. chim. **63**, 5 (1944).

36. HOSEMANN, R.: Acta Cryst. **4**, 520 (1951).

37. POROD, G.: Acta Phys. Austriaca **3**, 66 (1949).

38. BOLDUAN, O. E. A., and R. S. BEAR: J. Appl. Phys. **22**, 191 (1951); J. Polymer Sci. **6**, 271 (1951).

39. EWALD, P. P.: Proc. Phys. Soc. (Lond.) **52**, 167 (1940).

40. HOSEMANN, R.: Z. Physik **128**, 1, 465 (1950).

40a. BONART, R., u. R. HOSEMANN: Symposion über Makromoleküle. Wiesbaden 1959.

41. HOSEMANN, R., R. BONART u. G. SCHOKNECHT: Z. Physik **146**, 588 (1956).

42. JANESCHITZ-KRIEGL, H., O. KRATKY u. G. POROD: Z. Elektrochem. 56, 146 (1952).
43. KRATKY, O., u. H. SEMBACH: Angew. Chem. 67, 603 (1955).
44. — G. POROD u. A. SEKORA: Monatsh. Chem. 85, 1176 (1954). — I. PILZ u. A. SEKORA: Z. Naturforsch. 10b, 510 (1955); Z. Naturforsch. 11b, 10 (1956); Trans. Faraday Soc. 52, 558 (1956).
45. MACARTHUR, I., and B. PATNAIK: Proc. Leeds Phil. Lit. Soc. 5, 254 (1949).
46. HESS, K., u. H. KIESSIG: Z. physik. Chem. (A) 193, 196 (1944).
47. — — Kolloid-Z. 130, 10 (1953).
48. ARNETT, L. M., E. P. H. MEIBOHM and A. F. SMITH: J. Polymer Sci. 5, 737 (1950). — MEIBOHM, E. P. H., and A. F. SMITH: J. Polymer Sci. 7, 449 (1951).
49. FANKUCHEN, J., and H. MARK: J. Appl. Phys. 15, 364 (1944).
50. ZAHN, H., u. K. KOHLER: Kolloid-Z. 118, 115 (1950). — ZAHN, H.: Melliand Textilber. 32, 534 (1951). — ZAHN, H., u. U. WINTER: Kolloid-Z. 128, 142 (1952).
51. ROTHE, H.: Kolloid-Z. 151, 155 (1957); Faserforsch. u. Textiltech. 8, 244 (1957).
52. STATTON, W. O., and L. C. HOFFMAN: Nature (London) 176, 561 (1955).
53. KRATKY, O., A. SEKORA u. R. BREINER: Makromol. Chem. 22, 115 (1956). — KRATKY, O., u. R. BREINER: Makromol. Chem. 26, 92 (1957).
54. CLARK, G. L., and E. A. PARKER: Science 85, 203 (1937).
55. KIESSIG, H.: Papier 12, 117 (1958).
56. STATTON, W. O.: J. Polymer Sci. 22, 385 (1956).
57. KRATKY, O., u. A. SEKORA: Z. Naturforsch. 9b, 505 (1954).
58. SELLA, C.: Symp. Int. Chim. Macromol. Prag 1957, Mitt. Nr. 173; Compt. rend. 248, 1819 (1959); Symp. Makromol. Mitt. I B 7. Wiesbaden 1959.
59. MANDELKERN, L., C. R. WORTHINGTON and A. S. POSNER: Science 127, 1052 (1958).
59a. POSNER, A. S., L. MANDELKERN, C. R. WORTHINGTON and A. F. DIORIO: J. Appl. Phys. 31, 536 (1960).
60. STATTON, W. O., and G. M. GODARD: J. Appl. Phys. 28, 1111 (1957).
61. HENDUS, H.: Kolloid-Z. 165, 32 (1959).
62. FISCHER, E. W.: Z. Naturforsch. 12a, 753 (1957); 2 S.
63. KELLER, A.: Phil. Mag. 2, 1171 (1957).
64. TILL, P. H.: J. Polymer Sci. 24, 301 (1957).
65. STORKS, K. H.: J. Am. Chem. Soc. 60, 1753 (1957).
66. FRANK, F. C., A. KELLER and A. O'CONNOR: Phil. Mag. 4, 200 (1959).
67. STUART, H. A.: Kolloid-Z. 165, 3 (1959). — EPPE, R., E. W. FISCHER and H. A. STUART: J. Polymer Sci. 34, 721 (1959).
68. KELLER, A., and A. O'CONNOR: Nature (London) 180, 1289 (1957).
69. — — Faraday Soc. Disc. 25, 114 (1958); Symp. Makromol. Mitt. I B 1. Wiesbaden 1959.
70. ZAHN, H.: Symp. Makromol. Mitt. I B 8. Wiesbaden 1959.
71. HESS, K., u. H. MAHL: Naturwiss. 41, 86 (1954). — HESS, K., H. MAHL u. E. GÜTTER: Kolloid-Z. 155, 1 (1957).
72. STATTON, W. O.: J. Polymer Sci. 41, 143 (1959).
72a. — and P. H. GEIL: J. Appl. Polymer Sci. 3, 257 (1960).
73. HOSEMANN, R.: Kolloid-Z. 125, 149 (1952).
74. BONART, R.: Z. Krist., Mineral., Petrog. 109, 298, 309 (1957). — BONART, R., u. R. HOSEMANN: Z. Elektrochem. 64, 314 (1960).
75. KELLER, A.: Kolloid-Z. 165, 34 (1959).
76. MÜLLER, F. H.: Kolloid-Z. 108, 233 (1944); 165, 38 (1959). — FRANK, F. C.: Faraday Soc. Disc. 25, 208 (1958).

77. Peterlin, A., u. E. W. Fischer: Z. Physik 159, 272 (1960). — Fischer, E. W.: Z. Naturforsch. 14a, 584 (1959). — Peterlin, A., u. E. W. Fischer: Symp. Makromol. Mitt. I B 3. Wiesbaden 1959.
78. Wallner, L. G.: Monatsh. Chem. 79, 86, 279 (1948).
79. Hengstenberg, J., u. H. Mark: Z. Krist., Mineral., Petrog. 69, 271 (1928).
80. Kratky, O., A. Sekora u. R. Treer: Z. Elektrochem. 48, 587 (1942). — Heyn, A. N. J.: Nature (London) 172, 1000 (1953).
81. Heyn, A. N. J.: J. Am. Chem. Soc. 72, 2284, 5768 (1950).
82. Heikens, D., P. H. Hermans, P. F. van Velden and A. Weidinger: J. Polymer Sci. 11, 433 (1953). — Hermans, P. H., and A. Weidinger: J. Polymer Sci. 14, 397 (1954); Makromol. Chem. 13, 30 (1954).
83. Heyn, A. N. J.: Textile Research J. 23, 782 (1953); J. appl. Physics 26, 1113 (1955).
84. Fournet, G., and P. Antzenberger: Compt. rend. 236, 394 (1953).
85. Kratky, O., u. G. Porod: Die Physik der Hochpolymeren. Hrsg. von H. A. Stuart. III. Band, S. 212. Berlin-Göttingen-Heidelberg: Springer-Verlag 1955.
86. Kratky, O., u. A. Wurster: Z. Elektrochem. 50, 249 (1944). — Janeschitz-Kriegl, H., u. O. Kratky: Z. Elektrochem. 57, 42 (1953).
87. Stern, F.: Trans. Faraday Soc. 51, 430 (1955).
88. Hermans, P. H., D. Heikens and A. Weidinger: J. Polymer Sci. 35, 145 (1959). — Heikens, D.: J. Polymer Sci. 35, 139 (1959). — Hermans, P. H. and A. Weidinger: Makromol. Chem. 39, 67 (1960).

Fortschr. Hochpolym.-Forsch., Bd. 2, S. 401—441 (1961)

Mechanism of Acrylonitrile Polymerization

By

W. M. Thomas

Contribution from the Chemical Research Department, Central Research Division, American Cyanamid Company, Stamford, Connecticut

With 4 Figures

Contents

	Page
1. Introduction	401
2. Free-Radical Polymerization in Homogeneous Solution	402
3. Free-Radical Bulk Polymerization	409
4. Heterogeneous Free Radical Polymerization in the Presence of Organic Diluents and Additives	417
5. Polymerization in Aqueous Suspension	422
6. Anionic Polymerization	428
References	435

1. Introduction

This review deals with current ideas on the mechanisms operative in acrylonitrile polymerization. The topic is of importance in its own right and also because the study of acrylonitrile has cast light on heterogeneous polymerizations in general. It is an active field of research and the interpretations are still controversial. We shall look first at free-radical polymerization in homogeneous solution, where the monomer behaves in a more or less classical fashion. Next we shall consider the complications that arise where the monomer is at least partially soluble in the reaction medium but where the polymer precipitates. These conditions are encountered in bulk polymerization and in most aqueous or organic diluents. Finally we shall examine the less extensive literature on anionic polymerization and show important differences between the radical and the ionic processes.

The general literature on acrylonitrile monomer, its reactions, its polymerization and the technical applications of its polymers have been summarized in a recent book listing 1454 references (1). These topics will not be discussed here except as they bear on polymerization mechanisms. Copolymerization is mentioned only as it throws light on homopolymer formation. The present review covers the literature up to about January, 1960.

Before discussing details of the polymerization process, we may consider the nature of the polymer itself. Ordinary polyacrylonitrile appears to have the normal head-to-tail structure. Polymer made in homogeneous solution or in the presence of chain transfer agents is probably essentially unbranched. Polymer made in aqueous systems at temperatures below about 50° is very little branched (41, 67, 98). High molecular weight fractions of polymer prepared in aqueous systems appear to be somewhat branched (98, 111), especially when made above 50° and when active initiators like persulfate are used. Evidence for other structural modifications (cyanoethylation and ketene imine formation) is discussed in appropriate sections below. Recently two groups of investigators (138, 124) have begun the preparation and study of low molecular weight oligomers and analogs.

A large number of papers has dealt with the question of molecular weight and molecular weight distribution in polyacrylonitrile. Recent discussions are those of ONYON (108), KRIGBAUM and KOTLIAR (90), BOOTH and BEASON (33), BAMFORD, JENKINS, JOHNSTON and WHITE (19) and KOBAYASHI (87). The intrinsic viscosity vs. molecular weight relationship of CLELAND and STOCKMAYER (41) is probably as well supported as any. Fortunately, much of the material in this review does not depend heavily on detailed knowledge of molecular weights or of their distributions.

Finally, the question of stereoregularity and crystallinity must be mentioned. The polymer has often been called crystalline on the basis of X-ray diffraction (37, 123) solubility (33), spherulite formation (71) and other characteristics. Some of the confusion arises from differing definitions of what constitutes "crystallinity". The best interpretation seems to be that ordinary polyacrylonitrile can exhibit good lateral order (packing in directions perpendicular to the chain axis) but except in the most unusual circumstances (106) it is poorly ordered along the chain axis. The possibility of syndiotactic configuration has been discussed by several writers (2, 10, 90). Probably normal polymer is not completely atactic, but the extent of regularity has not been measured reliably. ARCUS and BOSE (2) were not able to induce crystallinity by treating polymer with bases to invert $>$CHCN groups. For a very recent discussion of crystallinity, glass temperature and chain conformation the reader is referred to the paper of KRIGBAUM and TOKITA (91).

2. Free-Radical Polymerization in Homogeneous Solution

The main features of acrylonitrile polymerization in solution have been clarified through a number of investigations in the past five years. The general conclusion to be drawn from this work is that in homogeneous systems acrylonitrile is a fairly typical vinyl monomer. Although this statement may not appear significant in itself, it becomes important when viewed in context with results on heterogeneous systems. The heterogeneous work, to be described in a later section, preceeded the solution work in the historical development of the subject and provided much of the impetus for undertaking solution studies.

In this review the elementary steps and their velocity coefficients are defined as follows:

Rate of Initiation $= R_i$
Rate of Polymerization $= R_p$
Initiator $= I$
Monomer $= M$
Solvent $= S$
Dead Polymer $= X$
Growing Polymer $= M \cdot$
Catalyst Efficiency $= f$
Degree of Polymerization $= P$

Radical formation

$$I \qquad\qquad \rightarrow 2\, I \cdot \qquad d\,[I\cdot]/dt = 2\,k_1\,[I]$$

Initiation

$$I \cdot = M \quad \rightarrow M \cdot \qquad R_i = 2\,f\,k_1\,[I]$$
(may depend on M at low f)

Propagation

$$M \cdot + M \quad \rightarrow M \cdot \qquad -d\,[M]/dt = k_2\,[M\cdot]\,[M]$$

Transfer to solvent

$$M \cdot + S \quad \rightarrow X + S \cdot \quad -d\,[M\cdot]/dt = k_3\,[M\cdot]\,[S]$$

Termination by combination

$$2\,M \cdot \qquad \rightarrow X \qquad -d\,[M\cdot]/dt = 2\,k_4\,[M\cdot]^2$$

Termination by metal ion

$$M \cdot + \mathrm{Fe}^{+3} \rightarrow X + \mathrm{Fe}^{+2} - d\,[M\cdot]/dt = k_5\,[M\cdot]\,[\mathrm{Fe}^{+3}]$$

Square brackets denote molar concentrations. Units are moles, liters, seconds, degrees centigrade and kilocalories per mole. E_n refers to the energy of activation for the step in which k_n is the velocity coefficient, the latter term being used to avoid reference to variable rate "constants". Since the polymerization rate, R_p, is commonly

$$R_p = -d\,[M]/dt = k_2(k_1 f/k_4)^{1/2}\,[I]^{1/2}\,[M]$$

it follows that the over-all activation energy, E_0, is in this instance

$$E_0 = E_2 + 1/2 E_1 - 1/2 E_4 .$$

The ratio k_3/k_2 is the chain transfer constant, C_s.

Much of the early literature on polymerization in homogeneous solution is cited by Ulbricht (*135*) and by Srinivasan and Santappa (*122*). Only a few liquids are suitable solvents for the polymer, and a greater part of the mechanism work has been concerned with N,N-dimethylformamide, generally abbreviated as DMF. In this review we describe the reaction in DMF initiated by an azo compound and then go to results with other systems for comparison.

Current views on polymerization of acrylonitrile in homogeneous solution are illustrated by a description of the reaction in N,N-dimethylformamide (DMF) as initiated by azobisisobutyronitrile (AIBN) at about 50 to 60°. Primary radicals from the decomposition of AIBN react with monomer to start a growing chain. About one-half of the primary radicals are effective, the others being lost in side reactions not leading to polymer. Bevington and Eaves (*32*) estimated initiator efficiency by use of AIBN labelled with C-14, whereas Bamford, Jenkins and Johnson (*13*) used the FeCl$_3$ termination technique. Both of these methods require that the rate of AIBN decomposition be known, and the numerical value of this rate has undergone a number of revisions that require recalculation of efficiency results. From recently proposed rate expressions for AIBN decomposition at 60° (*22, 136*) one calculates an efficiency of about 40% by the tracer technique and 60—65% by the FeCl$_3$ method.

It has been deduced from the order of reaction with respect to monomer that the rate of initiation is not independent of monomer (*122*). One could assume that initiator and monomer may form a complex from which the actual initiator is derived, or else that the AIBN and its radical products are in equilibrium within a cage of solvent. Although kinetic data are consistent with such assumptions, it appears that the results can be explained reasonably on the grounds of termination by primary radicals (*21*). The order of reaction with respect to AIBN initiator in nearly all of these studies is the expected 0.50 or possibly slightly higher (0.55 to 0.59).

Chain growth continues at a rate dependent on the concentrations of monomer $[M]$ and of active sites $[M\cdot]$. Monomer exponents in the range 1.3 to 1.5 or higher had been observed (*110, 123, 127*) especially at low $[M]$, but first order dependence has now been established over a broad range of $[M]$ (*21*). A stationary level of $[M\cdot]$ is reached rapidly and is typically of the order of 10^{-8} molar. Chains grow rapidly by successive monomer additions until the polymer chain is terminated by transfer or by reaction with another radical. The rate constant for propagation (k_2) at 60° in DMF is 1960 $m^{-1}ls^{-1}$ (*16*), which is a comparatively high value [see Table 1 and ref. (*16*)]. On the other hand it is only about one-tenth of that found for acrylonitrile in aqueous systems (Table 6) but is much above the range proposed (24) for post-polymerization. The ratio $k_2/(2\,k_4)^{1/2}$ has been assigned a variety of values, the most reliable of which appears to be 0.081 (21, 32). This corresponds to $k_2/k_4^{1/2} = 0.11$ in our notation. The value of $E_2 - \frac{1}{2}\,E_4$ has been given (122) as 7.6 kcal. per mole.

Termination is a bimolecular process involving the growing chain and a second radical, usually another growing chain. Termination by cyanopropyl radicals from AIBN takes place to some extent (27). It had been thought that the termination step involved disproportionation into an unsaturated polymer and a saturated polymer, but BAMFORD, JENKINS and JOHNSTON (20) have presented strong evidence that the two growing chains couple to give an inactive polymer molecule. In the first place, the relationship between degree of polymerization and initiator concentration is that required for coupling. Secondly, when polymer is prepared with initiator containing functional groups, it is possible to link chains subsequently through these end groups. The increase in solution viscosity on end-linking is consistent with coupling rather than with disproportionation. In this respect acrylonitrile behaves in an unusual manner, since most other vinyl polymerizations terminate by disproportionation. The work of BEVINGTON and EAVES (32) generally confirms the combination mechanism. If termination were by disproportionation and transfer to solvent occurred as has been supposed, then the number of initiator fragments per molecule would be less than they found. On the other hand, soluble graft polymers have been made at higher temperatures (100°—135°), so termination cannot have been largely by combination in these cases. Either the mechanism changes at higher temperatures or else transfer reactions prevent crosslinking [see discussion after reference (14)].

Table 1. *Velocity Coefficients for Acrylonitrile and Some Other Monomers in Homogeneous Systems*

	k_2	k_4 ($\times 10^{-7}$)
Acrylonitrile (in DMF) . .	1960	39.1
Vinyl acetate.	3700	7.4
Methyl acrylate.	2090	0.47
Methyl methacrylate . . .	367	0.93
Styrene	176	3.6

Data are for 60° C and non-aqueous systems. Except for acrylonitrile they are from the summary by FLORY (54). In BAMFORD's notation k_4 is double this amount.

Depending on the ratio of DMF to monomer, chains are stopped more or less frequently by abstraction of a hydrogen atom from the DMF solvent. The point of attack on DMF was thought (109, 122, 127, 135) to be as in A below, but a paper by CAMPBELL (36) suggested that the N-methyl group is attacked when BF_3 is present and BAMFORD and

27*

White (*18*) have now shown that the DMF radical is indeed the one of structure B:

$$\overset{\overset{\displaystyle O}{\parallel}}{\cdot C} N(CH_3)_2 \quad (A) \qquad\qquad HC\overset{\overset{\displaystyle O}{\parallel}}{N}\underset{CH_2\cdot}{\overset{CH_3}{<}} \quad (B)$$

Radical B may of course result also from attack by radicals from AIBN. In that event or in the event of transfer from the polymer radical to solvent, a new polymer chain will be started by B and this chain will have no AIBN fragment at that end. Coupling of two B radicals is assumed to be a very infrequent event. Transfer to monomer, to polymer or to initiator are infrequent also. The transfer constants are approximately 1×10^{-5} for transfer to monomer at 25° (*108*) and 2.7×10^{-4} for transfer to DMF at 50° (*135*). Srinivasan and Santappa (*126*) suggest that the values for DMF are 1.0×10^{-4} at 50° and 2.4×10^{-4} at 60°, although this requires an unexpectedly high activation energy of 18 kcal per mole for transfer. Hydrated DMF appears to be relatively inert toward transfer (*84*).

The question then arises as to how rapidly radical B reinitiates a growing chain. Thomas, Gleason and Pellon (*127*) and then Ulbricht (*135*) observed that the polymerization rate in DMF is slower than in some other solvents and that in the polymerization rate expression the monomer exponent is fractionally higher than unity (*109, 122*). These and molecular weight considerations led them to conclude that the rate of reinitiation is relatively slow [see also ref. (*109*)]. Kinetic expressions were derived that reduce to the conventional equation only in the limit of high monomer concentration. Rapid cross-termination as in copolymerization may be invoked also (*22, 109*). Jenkins (*83*) on the other hand has concluded that explanations of this type [see also Burnett and Loan (*35*)] are not likely to be general in polymerization kinetics. Bamford, Jenkins and Johnston (*21*) showed recently (as mentioned above) that the $[M]^{1.0}$ relationship is obeyed at high $[M]$ and that deviations at low $[M]$ can be explained by assuming that a significant termination step is reaction of primary radicals (from AIBN) with the growing chain. Further work in solvents other than DMF will be helpful at this point.

Post effects have not been examined in detail in homogeneous systems. Onyon (*109*) attempted to use the viscometric method, but the data did not vary in accordance with theory and did not allow accurate determination of individual rate constants.

The general picture outlined so far is presumed to be valid for other solvents, initiators and temperatures, but the evidence is much less extensive. The choice of solvents is quite limited, and at high $[M]$ these liquid mixtures lose their solvent properties. This limit is about 3 molar for DMF at 25° and about 5 or 6 molar at 60°. Polymerization in ethylene carbonate and in 4-butyrolactone is considerably more rapid than in DMF (127, 135), supporting the idea that radicals derived from DMF are relatively inactive. ULBRICHT (135) concluded that the transfer constants to solvent at 50° are

$$\begin{aligned}
\text{Ethylene carbonate} & \quad . \quad . \quad 3.9 \times 10^{-5} \\
\text{Butyrolactone} & \quad . \quad . \quad . \quad . \quad 7.4 \times 10^{-5} \\
\text{Dimethylformamide} & \quad . \quad . \quad 27.0 \times 10^{-5}
\end{aligned}$$

CLELAND and STOCKMAYER (41) used succinonitrile as one of their solvents, but is has been suggested since that polymerization may not have been straightforward in that medium (19).

Polyacrylonitrile is soluble in a variety of concentrated aqueous salt solutions, and these solutions may be used in acrylic fiber manufacture. On several occasions it has been proposed that polymer might be formed directly in such solutions, thus avoiding the intermediate steps of separating and dissolving the polymer. One such reference is a recent patent by SCHMIDT (118). HUNYAR and GROBE (74) discussed the variables in their photoinitiation process, and GROBE and SPODE (65) studied polymerization as a function of the wave length of the light. MILLER (99) has preferred aqueous perchlorates in grafting experiments because the solvent is not a chain transfer medium and because it is not attacked by oxidizing catalysts. Extensive mechanism studies have not been carried out in these aqueous solvents.

ULBRICHT (135) calculated initiator efficiencies as follows, using AIBN. He concluded that disproportionation is the more likely, but BAMFORD, JENKINS and JOHNSTON (20) used ethyl-

Assumed Termination Mechanism	f	
	Ethylene Carbonate	Butyro-lactone
Combination	0.96	0.83
Disproportionation .	0.48	0.42

ene carbonate as well as DMF in their experiments and concluded that combination is operative in both solvents. If so, the efficiencies of ULBRICHT seem unreasonably high.

Experiments in which initiators other than AIBN are used do not indicate any unusual effects. Polymerization rate is nearly always proportional to the square root of the initiator concentration or at least to a value between 0.5 and 0.6. ULBRICHT (135) established the square root law for ammonium persulfate, AIBN and benzoyl peroxide in DMF, the rates being fastest with persulfate, slower with AIBN and slowest with the peroxide. One expects that persulfate and peroxide will be more active than AIBN in abstracting hydrogen from other components of the systems. Other initiators have been used, including UV with di-t-butyl peroxide (*109*) and either Ra or Co-60 (*28*). Results are in general conformity to those observed with AIBN. BENSASSON and PREVOT-BERNAS (*28*) found that with ionizing radiation the intensity exponent in 70 mole percent DMF is 0.55, which is substantially the same as the exponent for AIBN.

BARSON, BEVINGTON and EAVES have looked into the initiation reaction with azobisisobutyronitrile and with benzoyl peroxide, both in dimethylformamide (*23*). With AIBN they reported normal initiation. With the peroxide there appeared to be complications, due they thought to hydrogen abstraction from monomer or polymer. By using peroxide labelled in the ring or in the carbonyl group with C-14 they measured the relative importance of the two steps:

$$C_6H_5COO \cdot \rightarrow C_6H_5 \cdot + CO_2 \tag{A}$$

$$C_6H_5COO \cdot + CH_2{=}CHCN \rightarrow C_6H_5COOCH_2\dot{C}HCN \tag{B}$$

and found that the ratio of rate B/rate A is not less than 8 m l^{-1} at 60°. This ratio varies from monomer to monomer and is greater for styrene than for acrylonitrile. Thus the first unit in a styrene copolymer would be $C_6H_5COOCH_2\dot{C}HC_6H_5$ rather than a unit containing acrylonitrile.

Small quantities of inhibitors and retarders give the expected effects. Most of the solvents for polyacrylonitrile are difficult to purify, and some of the literature is based on results obtained with solvents containing traces of amines and water. The growing chain is an electron acceptor and undergoes transfer reactions most readily with electron donors (*14*). Addition of small amounts of agents like CBr_4, which commonly serve as efficient chain transfer agents, is considerably less effective. On the other hand, triethyl amine is a very efficient transfer agent and this fact has been used as a basis for preparing block copolymers (*15, 22*).

When certain salts like $FeCl_3$ are present the major termination reaction becomes electron transfer from the growing chain, reducing Fe^{+3} to Fe^{+2}. The rate of Fe^{+2} formation can be identified with the rates of initiation and termination. BAMFORD and coworkers have used this principle to estimate initiator efficiencies and individual rate constants (*12, 13, 16, 19*). At 60° the rate constant (k_5) for ferric ion termination was estimated as 6533 l m^{-1} s^{-1} (*13*). In these experiments the initiator efficiency f was calculated to be in the range 0.73 to 0.79, which now appears high.

Other salts probably have similar effects but have not been studied extensively. Lithium salts have little influence on the termination reaction but can increase the propagation rate, presumably by complexing the growing nitrile radical (*14*). Anions complex with the growing chain in the order $Cl > NO_3 > ClO_4$. Complex formation between salt and monomer appears unimportant, but salt-initiated anionic polymerization has been suggested (*14*). These salt effects are by no means limited to acrylonitrile polymerization but have been observed with acrylamide and with other monomers.

Distribution of molecular weights in these homogeneous systems has been uncertain. A recent discussion and review of this topic is that of BAMFORD, JENKINS, JOHNSTON, and WHITE (*19*). They assume a simple exponential distribution and consider both transfer and bimolecular termination. PEEBLES (*112*) observed conventional polymerization kinetics but he concluded from light scattering results that there are deviations from normal behavior in the higher molecular weight fractions. He associated these with the same mechanism that causes microgel.

3. Free-Radical Bulk Polymerization

It might be supposed that the simplest way to study acrylonitrile polymerization would be to follow the reactions of pure monomer. In fact, however, this system is very complex and even now after many investigations the details are not completely understood. This is because separation of polymer in an essentially unswollen state leads to kinetic results that have been interpreted in a variety of ways. The subject has been treated recently in a book by BAMFORD, BARB, JENKINS and ONYON (*17*).

Pure monomer is quite stable to heat, although the cyclic dimer

$$CH_2CHCN$$
$$CH_2CHCN$$

is formed at temperatures in the range of 200 to 300° (*44*). For most purposes uncatalyzed thermal polymerization can thus be disregarded. A further simplification arises from the fact that depolymerization is unimportant under usual conditions.

Polymerization is initiated readily by light, by azo and peroxy compounds, by ionizing radiation, or in general, by any source of free radicals. If a tube containing pure acrylonitrile together with a trace of catalyst is kept at a moderate temperature there is a period of a few seconds to minutes during which the solution remains clear and while traces of inhibitors such as oxygen are being consumed. At this stage the monomer contains particles roughly 100 to 200 Å across. They are slightly flattened spheres not large enough or numerous enough to cause turbidity (*130*).

Sudden appearance of a haze signals the beginning of the main polymerization reaction. Ordinarily the haze does not extend uniformly through the tube but appears as a cloud along the axis of the tube beginning near the bottom. Turbidity increases rapidly, and if the tube is shaken, discrete particles can be seen with the unaided eye. If the tube is kept still a clear layer may persist for a short time. After about 0.2 to 0.3% of the monomer is converted to polymer the fine suspension begins to coagulate to a curdy precipitate and the slurry gradually increases in thickness. By the time half of the monomer has been converted to polymer the reaction mixture is a hard, coarse, white solid. Particles

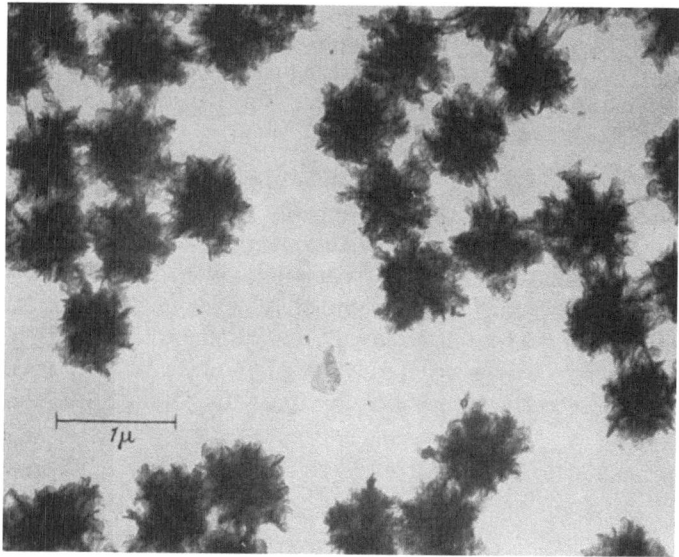

Fig. 1. *Polyacrylonitrile.* Bulk polymerization at 60°; (*127*)

are fairly densely packed aggregates porous to nitrogen. They are several thousand Angstroms across and comprise many smaller particles in the range of a few hundred Angstroms to 1000 Å or more (*126*). Each of the smaller particles contains many polymer molecules. Fig. 1 shows aggregates made at 60°; at 30° they are less uniform (*130*).

A large amount of heat (17 kcal/mole) is given off during polymerization (*131*). The character of the slurry and the speed of reaction are such that this heat is not easily dissipated. Unless the sample size is small and the rate of initiation is low, the temperature is likely to rise rapidly and lead to an uncontrolled reaction. These characteristics make bulk polymerization unattractive except for laboratory experimentation.

First detailed studies appeared in the 1940's, notably those of KERN and FERNOW (*86*) and of KONINGSBERGER and SALOMON (*89*). Our knowledge of the mechanism of bulk polymerization is based largely on kinetic work of the present decade. Results turned out to be quite different from those generally obtained with vinyl monomers or even with acrylonitrile itself in homogeneous solution. The more striking differences are shown in Table 2 and are discussed in turn below.

A typical conversion-time curve exhibits a short period of acceleration followed by a nearly constant rate up to 50% conversion or more, then followed in turn by a diminishing rate. The acceleration period is sometimes reported to persist to quite high conversion. Although some

Table 2. *Comparison of Bulk with Solution Polymerization*

	Heterogeneous (Bulk)	Homogeneous (Solution)
Acceleration period	Present	Absent
Monomer exponent in rate expression	Not observed	1.0 over most of range
Initiator exponent in rate expression	0.7 – 0.8	0.5 – 0.6
Energy of activation	Individual values not known; some appear to be high	Normal
Velocity coefficients	Variable and not known with certainty	Normal (rather high)
Post polymerization	Important; shows unusual features such as fast reaction	Moderate (does not fit usual expression)
Mol. wt. vs. temperature	Maximum at 60° – 60°	Normal inverse relationship with chemical initiation
Mol. wt. vs. initiator	$I^{-0.3}$	$I^{-0.5}$
Radical concentration		
During polymerization	ca. 10^{-4} to 10^{-5} M	Normal, ca. 10^{-8} M
At end of reaction	Appreciable fraction of radicals trapped	Decays rapidly to zero

workers (*122*, *27*) have recorded the usual square root dependence on initiator, at least at the very beginning, an initiator exponent of about 0.8 is observed in the constant rate period. This type of dependence holds regardless of whether the initiation is by azo compounds, peroxides, sensitized ultraviolet light or ionizing radiation. With high intensity X-rays the exponent appears to drop somewhat (*31*, *39*), although accurate dosimetry is more difficult in this range.

Originally it was thought that the capture of radicals by acrylonitrile is so rapid that the catalyst efficiency is near unity (*3*). In view of more recent results in homogeneous solution, it is perhaps more reasonable to assume that the efficiency is likely to be in the range 0.5 to 0.7 (*32*).

In ordinary vinyl polymerizations, molecular weight varies inversely with the square root of the initiator concentration. In the case of acrylonitrile at 40°, intrinsic viscosity varies as $I^{-0.2}$ which implies that molecular weight varies as $I^{-0.3}$ (*126*).

If the source of free radicals is removed, as can be done conveniently with UV initiation or with ionizing radiation, polymer continues to form although at a greatly reduced rate. This post-effect is substantial and may persist for some time. According to MAGAT (93) the rate may not decay to 1% of its initial value in 100 days. In one experiment conversion was 7% at the end of irradiation and had increased to 11% at the end of the post-reaction (93). The magnitude of the post-effect depends on the extent of the previous photo-reaction and also on the physical state of the polymer (7, 29). It will be recalled that in ordinary homogeneous polymerization the post-effect does not depend on the length of the prior illumination except for very short exposures. For relatively long post-effect times the post-conversion is a logarithmic function of time, but the exact functional dependence is a matter of some uncertainty (29). BENSASSON and BERNAS compared polymer before and after post-polymerization and found that the latter was less readily soluble, gave a lower intrinsic viscosity and a higher HUGGINS constant (29). They suggested that post-polymerization may involve branching. BAWN, HOBIN and McGARRY (24) have irradiated monomer at 60° and have observed the post effect at that temperature. They showed that the radical concentration is of the order of 10^{-5} molar and that radicals may be kept without loss for a week at −80°. Assuming that k_4 did not vary (a procedure criticized in the ensuing discussion) they estimated k_2 as about 0.3 $1\,m^{-1}\,s^{-1}$ and k_4 as 24 to 51 $1\,m^{-1}\,s^{-1}$. These are extraordinarily low values.

An interesting consequence of the properties of acrylonitrile is that it enabled BAMFORD and JENKINS (9) to observe for the first time a thermal after-effect. A catalyzed polymerization was carried to about 4% conversion at 60°, chilled to −80° and then continued at 25°. The 25° rate was about twice the usual 25° rate, indicating that the higher radical concentration at 60° had been preserved at least in part.

The temperature dependence of bulk polymerization shows some unusual features. BAMFORD and JENKINS measured initial rates and obtained an activation energy of about 28 kcal per mole (7). THOMAS and PELLON found 35 to 37 kcal for the constant rate period up to about 50% conversion (126). These authors used azo and peroxy initiators in the range 30 to 60°. More recently BENSASSON and BERNAS (29) initiated polymerization with gamma rays and obtained about 5 kcal between 0° and 20° and about 15 kcal between 20° and 50°. One expects a drop of about 15 kcal in going from azo or peroxy initiators to initiation by radiation, since splitting an initiator molecule is avoided. Normal values for other monomers in radiation work are 4 to 7 kcal. If irradiation is at 20° and the post-effect is measured at 40° to 60°, the activation energy with respect to initial rates is very high (35 kcal). These figures cannot be rationalized completely because the values for the individual steps can only be guessed. It appears probable that separation of solid polymer reduces both propagation and termination rates although not to the same relative extents. At temperatures near 60° there is a greater effect on termination than on propagation resulting in a maximum in molecular weight.

Recent work with ionizing radiation suggests that the polymerization mechanism at low temperatures may well be anionic (95) rather than free radical. CHEN, COLTHUP, DEICHERT and WEBB (40) have shown that polymer made with X-rays at —78.5° exhibits an infrared band at 2030 cm^{-1} characteristic of ketene imines ($>$C=C=N–). The mechanism by which this structure forms is not clear but it does not seem to be introduced by action of radiation on the finished polymer. According to GRASSIE and McNEILL (64) ketene imines are formed in polymethacrylonitrile by reaction of the growing radical in the form —CH$_2$–C=C=N.

$$\overset{|}{\text{CH}_3}$$

BENSASSON (30) has polymerized acrylonitrile even in the solid state at —196° and finds that it polymerizes more rapidly at —196° than in the liquid at —78°. Very recently SOBUE and TABATA (121) have investigated polymerization induced in bulk by ionizing radiation over the range + 15 to —196°. They suggest that both ionic and radical polymerization may occur.

Another unusual temperature effect is observed in comparing the intrinsic viscosity of polymer prepared with chemical initiators at a series of temperatures from 25° to 100° (7, 126). A distinct maximum appears at a temperature of preparation in the 50° to 60° range, whereas one expects a monotonic decline with temperature. Typical figures are the following (126) (see table 3). Other temperature effects are mentioned in the discussion of the fast reaction after effect.

Not much is known about molecular weight distribution in bulk polymers. An intrinsic viscosity of 10 deciliters per gram corresponds to a weight average molecular

Table 3. *Intrinsic Viscosity vs. Temperature Bulk Polymerization with 0.008 Molar Benzoyl Peroxide*

Polym. Temp. °C	Intrinsic Viscosity in DMF at 25°
30	11.8
40	12.7
50	13.1
60	11.0

weight of perhaps a million, although relationships in this range are not well established. It is generally assumed that molecular weight distribution is quite broad and that in some cases there may be microgel. Although branching is not a frequent reaction, it may be favored somewhat by heterogeneous conditions. Branching is more probable at high temperatures and the most branches will be found in the largest molecules.

Under a given set of conditions intrinsic viscosity changes only moderately with conversion, at least in the 5 to 55% range. Some data of Thomas and Pellon (126) illustrate this point (Table 4).

Some attempts have been made to estimate individual velocity coefficients, most recently by Bengough (27). In view of the complexities one must regard the results as tentative. If the physical state of the system does indeed alter velocity coefficients drastically, there is no single value but rather a whole spectrum related to the particular environment of each growing chain. One might expect that conditions in the very earliest stages of reaction would approximate those in homogeneous solution. If anything, the values for bulk propagation and termination are lower than in homogeneous solution (DMF) and very much lower than in water. The ratios $k_2/(2\,k_4)^{1/2}$ are considerably less affected, however.

These kinetic results have led to a variety of interpretations. Magat and associates have proposed kinetic schemes based on the idea that a steady state does not exist and that only small (primary) radicals can terminate polymer chains (52, 93). Bamford and Jenkins have criticized this concept of emulsion-type polymerization (11). They point out that if the source of initiation were removed no more small radicals would be formed and polymerization should continue indefinitely. They cite a photosensitized reaction at 60° in which the light was shut off at about 15% conversion, whereupon the rate fell to one-half of its original value in 60 seconds. Bamford and Jenkins point also to evidence from the fast reaction that argues against emphasis on termination between polymer radicals and small radicals.

Table 4. *Intrinsic Viscosity vs. Conversion*

% Conversion	Intrinsic Viscosity
5.4	10.0
9.4	11.2
14.4	11.1
27.4	12.0
35.0	11.8
39.6	11.8
54.4	12.4

Thomas and Pellon (126), extending a kinetic scheme of Bernstein and coworkers (31), explained their results on the basis of two distinct termination steps. One of these was considered to be normal bimolecular termination, and the other was regarded as a unimolecular process in which the growing chain becomes buried and is shielded from further growth. As polymerization progresses monomer is depleted more rapidly than catalyst is decomposed, and the liquid phase is enriched in catalyst. A kinetic scheme postulating a steady state and encompassing these considerations led to satisfactory agreement with the data, at least within the range of their experiments [although not necessarily under all conditions of later workers (115)].

Bamford and Jenkins (7, 8, 9, 10, 11) took exception to this treatment as being an undue simplification. They pointed out that both the propagation and termination coefficients are reduced by occlusion and that the extent of reduction is dependent on the physical state of the system, complete burial being relatively rare. Evidence presented below confirms the view the system is indeed complex, although it may be emphasized that velocity coefficients are not necessarily constant properties of a monomer even in systems that appear homogeneous on a macroscopic scale.

Until velocity coefficients and radical concentrations are known with greater certainty one cannot be sure how closely the true state of affairs is approximated by an algebraic treatment. Further effort to describe these heterogeneous systems by formal kinetics does not appear warranted at present. Progress is more likely to result from detailed investigations into the physical state of these systems. It seems quite possible that polymerization is occurring simultaneously on the particles and, because of slow precipitation, in the liquid phase as well. This would correspond to the situation described in a later section for aqueous polymerization.

Recent microscopical studies by THOMAS, THOMAS and DEICHERT (130) lend support to view that these systems share some features of emulsion polymerization. During the early stages of polymerization the particles grow as rather uniform aggregates (Fig. 1). The number of these particles is fairly constant and is in the range of 10^{11} to 10^{12} per cc. This is approaching the number found in typical emulsion systems.

The concentration of trapped radicals and the degree of occlusion (how deeply they are buried) have been studied extensively by BAMFORD and JENKINS (9). They determined the approximate radical concentration by swelling or dissolving polymer in solutions of diphenylpicryl hydrazyl (DPPH), a violet substance that becomes colorless on reaction with radicals. Nitrobenzene was generally used as the medium to dissolve the hydrazyl and to swell the polymer; β-propiolactone, a solvent for the polymer, was used also.

The concentration of trapped radicals in a typical experiment was estimated as about 3×10^{-5} mol/l of reaction mixture, or about 2×10^{16} radicals/ml. This concentration increased with the rate of photopolymerization, but the number of radicals that became trapped was a constant fraction of the total number generated. In these experiments that fraction was estimated at about 1% at 25°.

More recently, electron paramagnetic resonance has been used in place of DPPH to estimate radical concentrations (10, 80). Considering the uncertainties attendant on the use of DPPH the agreement is reasonably good. Under conditions where BAMFORD and JENKINS using DPPH found 5×10^{16} radicals per ml, INGRAM, SYMONS and TOWNSEND reported $3.5 \pm 0.05 \times 10^{17}$ with e.p.r. (80). The latter authors point out that polyacrylonitrile is far more efficient in occluding radicals than are any of the other compounds studied. The e.p.r. spectrum of polyacrylonitrile shows very little structure. Unfortunately, conclusions regarding polymerization mechanism are subject not only to the accuracy of e.p.r. measurements but also to estimates of catalyst efficiency and number average molecular weight.

If polymerization is carried to say 20% conversion at 25° and the reaction mixture is transferred to a 60° bath, there is a sudden fast reaction followed by the normal 60° rate (7, 9). The fast reaction is attributed to radicals trapped at 25° and liberated at 60°. Successive fast reactions can be observed by preparing polymer at 25, 60, and 70°. No further reaction occurs at 70°. From this fact, from the much greater rate at 60° than at 40° and from the fact that maximum molecular weight polymer is made at 50 to 60°, it is concluded that polymer undergoes marked changes in the range 40 to 60°. Polymer heated in alcohol at 60° contracts very slightly (7), but data are lacking on changes in the polymer in the 20° to 60° range and on quantitative aspects of swelling by monomer. The glass transition temperature measured by differential thermal analysis (85) is about 80°, though other techniques suggest higher temperatures (91, 73). The radical concentration was only slightly diminished by heating at 40°, but the degree of occlusion was probably increased. As the temperature and consequently the rate of photopolymerization were increased (15.8 to 44.0°) the rate of the fast reaction at 60° decreased, and it was concluded that the fraction of radicals trapped had decreased also. Experiments with e.p.r. suggest a smaller decrease as shown in Table 5 (81).

Table 5. *Trapped Radicals in Polyacrylonitrile*

Polymer Made at	Relative Radical Concentration
20°	8
40°	5
60°	2

When polymer was heated in the absence of monomer to 60° no decrease in radical concentration was observed over a long period of time. If oxygen is admitted to the e.p.r. tube at room temperature the resonance decays within 10 min to less than one-fifth of its value, but at −50° no change is observed in 24 hours.

If a swelling agent is added to the reaction mixture, a maximum rate is usually noted for some fairly low agent-to-monomer ratio. DMF, which is merely a swelling agent when mixed with monomer, gives a maximum rate at 10 mol-% in photoinitiated polymerization, at about 25 mol-% in polymerization catalyzed by benzoyl peroxide (9), and at about 30 mol-% with gamma rays (114), all near room temperature. On the other hand, as little as 10 mol-% of DMF reduces the rate at 60° by a factor of about 15. It decreases also the ratio of the fast reaction at 60° to the 25° rate.

Assuming that DMF is functioning chiefly as a swelling agent, rather than as a transfer agent and retarder, BAMFORD and JENKINS (9) interpret these results in terms of the degree of occlusion and its effect on the propagation and termination coefficients. With a small amount of swelling the effective propagation constant is increased. The number of radicals trapped may be somewhat diminished but those trapped are thought to be buried in a deeper mass of polymer. Mild swelling is thus comparable to heating at 40°, and such systems give the largest fast reactions on heating to 60°. Beyond a degree of swelling the termination rate is greatly enhanced (especially at higher temperatures) and the number of trapped radicals is very small. Reduction in the number of buried radicals far more than compensates for increased degree of occlusion. Thus the effect of occlusion is to decrease the termination coefficient and, to a lesser extent, the propagation coefficient. Either no occlusion or complete occlusion will prevent the fast reaction, of course. When polymer is warmed with DMF in the absence of monomer, partial solution takes place and the e.p.r. spectrum vanishes.

Copolymerization with small amounts of methyl acrylate produces results even more dramatic than those with DMF. As little as 0.03 mol-% is able to change the properties of aggregates enough to facilitate termination (9).

Not much is known about the detailed mechanism of occlusion. The process can be regarded as one of local monomer depletion in the vicinity of a growing chain. Coalescence of particles may be a more important factor than is the coiling of chains. Various pictures might be drawn, but necessary information on physical characteristics of the polymer is still lacking. For example, HAM (68) thinks it probable that in the early stages of polymerization monomer adds to the growing chain in a syndiotactic manner. He proposes that at the end of the "induction" period the polymer crystallizes and subsequently a more rapid helical, stereospecific polymerization occurs on the particles. The monomer is regarded as having sufficient solvating power to provide the necessary chain mobility. Until more data are available the situation will remain speculative. Information is needed about the formation of nuclei, polymer separation, coalescence of particles, polymerization loci, radical occlusion, swelling of particles and local monomer depletion.

4. Heterogeneous Free Radical Polymerization in the Presence of Organic Diluents and Additives

The general features of bulk polymerization are retained in systems to which non-solvents have been added. Microscopical studies (130) demonstrate that the structure of the particles can be varied from very dense to very loose by proper choice of diluent. Fig. 2 shows that polymer made in benzene is more densely packed than that made in undiluted monomer (Fig. 1). On the other hand Fig. 2 shows that polymer made in benzene-DMF mixtures may be very loosely packed. When the monomer is diluted to 20% by weight in acetone the number of particles formed is larger than in bulk and these particles are less densely aggregated.

Overall rates of polymerization are generally rather insensitive to the presence of diluent. DAS, CHATTERJEE and PALIT (50) compared rates at 50 mole percent concentration in a variety of liquids. In the range of their experiments, polymerization rate was nearly independent of the choice of diluent. BAMFORD, JENKINS and WHITE (22) point out that transfer agents reduce the mean degree of occlusion. Since the termination rate is increased under these conditions, the overall polymerization rate is reduced. The extent of polymer swelling will vary from one liquid to another (32, 130), and this obscures the interpretation to be made from limited data.

Generally there is a period of acceleration up to a few percent conversion, and this is often followed by a constant rate period extending to fairly high conversion. Kinetic data are considerably less extensive in these systems than in bulk, and interpretations are open to the criticisms that were cited for bulk work. The analogy to emulsion polymerization seems quite reasonable in systems containing organic diluents. Concurrent polymerization in both the liquid medium and the particles is a possibility. Monomer adsorption by polymer and extent of aggregation and swelling must be important factors.

Studies in benzene were reported by IMOTO and TAKATSUGI (77). With AIBN at 50° they found an initial acceleration followed by a constant rate period extending to 20% conversion or more. Rate in the monomer range below one molar varied as about $[M]^{1.6}$ and in the higher range as about $[M]^{3.0}$. A similar situation prevailed in respect to initiator, the exponent being 0.6 at low AIBN levels and 1.0 at higher concentrations. If DMF was present also, the lower figure was 0.8. An algebraic treatment employing spontaneous termination (unimolecular burial) and the steady state assumptions led to a rate expression more nearly in accord with their data than is the similarly derived expression of reference (126).

Experiments in toluene are included in the paper of SRINIVASAN and SANTAPPA (122). They worked with AIBN, chiefly at 50° and at 60°. The initiator exponents were 0.62 (50°) and 0.64 (60°) and the monomer exponent was about two. The authors concluded that both the initiation step and the termination-step are complex in these precipitating systems. They estimated that catalyst efficiency in toluene is only about 56% at

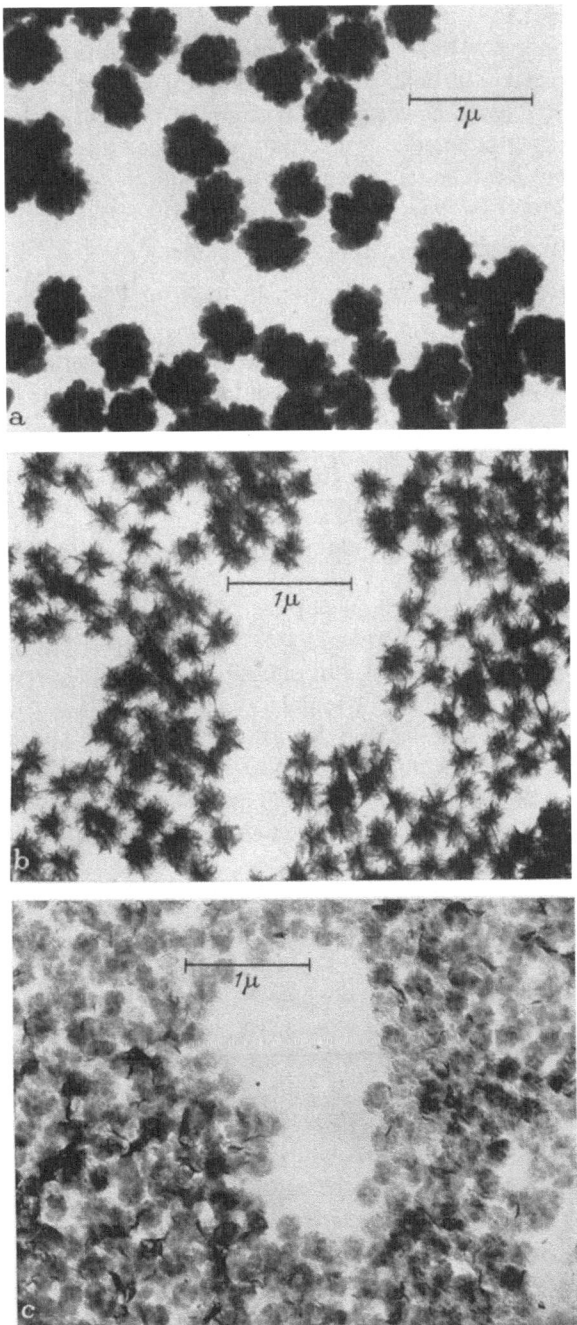

Fig. 2a—c. *Polyacrylonitrile (127)* Benzene-Dimethylformamide Media: 60° C; AIBN 3×10^{-3}, monomer 20% by volume. a) 100% Benzene; b) 3:1 Benzene : Dimethylformamide; c) 1:1 Benzene. Dimethylformamide

50°, and about 22% at 60°. Values of $k_2/(2\,k_4)^{1/2}$ are recorded along with the corresponding activation energies, and these quantities are compared for polymerization in bulk, in DMF and in toluene. The higher ratio of $k_2/(2\,k_4)^{1/2}$ in toluene as compared with DMF may mean a decrease in termination (k_4) in toluene. Since data in toluene are restricted to 50° and 60° nothing can be said about the maximum in $k_4/(2\,k_4)^{1/2}$ that these authors recorded for bulk polymerization or that others (7, 126) noted for molecular weight. Activation energies for $E_2 - \frac{1}{2}\,E_4$ were 8.0, 7.6 and 4.9 kcal/mole for polymerization in bulk, in DMF and in toluene, respectively.

NAKATSUKA and coworkers (104) measured polymerization rates in toluene with benzoyl peroxide at 30° and at 68°. Exponents in their rate expression were:

| Peroxide | 0.85 at 30°, | 1.0 at 68° |
| Monomer | 1.1 at 30°, | 2.0 at 68° |

A catalyst-monomer complex and spontaneous termination were postulated.

More complex catalysts have not been studied in detail. An exception is the work of IMOTO and TAKEMOTO (75) who investigated polymerization rates in benzene using a series of substituted benzoyl peroxides along and with dimethylaniline. They found a rough linear relationship between $\log (R/R_\mathrm{H})$ and σ, where R and R_H are rates with the substituted and unsubstituted peroxide and σ is the HAMMETT constant. The overall rate depended on the monomer to the first power and peroxide and amine each to the one-half power. They concluded tentatively that the benzoyloxy radical is the initiating species.

Systems containing monomer and DMF but no other diluent were discussed in the previous section, and it was noted that a maximum rate is associated with intermediate monomer-DMF compositions. IMOTO (79) observed a similar effect with monomer-benzene-DMF mixtures. A shallow minimum in polymerization rate was found at a low benzene level and then a very pronounced maximum at a higher benzene content. Molecular weight increased monotonically, however, with the benzene/DMF ratio. In these ternary systems the amount of DMF was not enough to keep the polymer in solution but considerable swelling must have occurred. THOMAS, THOMAS and DEICHERT (130) show electron photomicrographs of polymer made under these conditions.

NAKATSUKA (104) measured rates and molecular weights in mixtures of DMF and toluene at 30° to 50°. Benzoyl peroxide was the initiator. Molecular weight dropped as the temperature was raised and as the DMF content increased. Molecular weight distribution, for which the author relied on a methanol precipitation method, was surprisingly narrow. Polymerization rate at 30° was proportional to the peroxide concentration and to the square of the monomer concentration, just as had been found for toluene alone. A substantial after-effect was observed in toluene-DMF when acriflavine was used as a photosensitizer, and it was concluded that the radical lifetime near 20° is of the order of 10 min. To explain these results NAKATSUKA proposed that occlusion is important near 30°, the rate being controlled by the stage before occlusion and the molecular weight by the stage after occlusion. At higher temperatures occlusion was regarded as unimportant, but unimolecular termination was invoked in the kinetic scheme.

Using the same toluene-benzoyl peroxide system NAKATSUKA (105) measured polymerization rate and molecular weight as functions of temperature (40° and 58°) and of the concentration of three retarders: p-nitrophenol, 2,4-dinitrophenol and picric acid. Results were consistent with a kinetic scheme postulating (among other things) bimolecular initiation involving peroxide and monomer and spontaneous unimolecular termination of growing polymer chains.

OKAMURA, KATAGIRI and TAKEMOTO (107) added surface active agents to monomer containing benzoyl peroxide. Although the data do not permit strict comparisons, it appears that both polymerization rate and molecular weight were reduced by these agents. Variations were noted within each class of agent — non-ionic, cationic and anionic. When a polyglycol laurate was present, the dependence of rate on peroxide concentration dropped from 0.7 to approach 0.5. The authors felt that surface active agents prevent adhesion and growth of polymer particles and also prevent radical occlusion.

The radiotracer method for estimating efficiency of initiation was applied by BEVINGTON and EAVES (32) to polymerization in benzene and in carbon tetrachloride. Whereas they had calculated that about 47% of the radicals from AIBN initiate polymer chains in DMF solvent, efficiency in benzene was about 50% and in carbon tetrachloride about 30%. This low efficiency in carbon tetrachloride is attributed to attack of radicals from AIBN on the carbon tetrachloride solvent, especially at high concentrations of solvent. Chains initiated by secondary radicals derived in this way from the solvent would not be detected by the tracer method.

The most extensive chain transfer studies are those if DAS, CHATTERJEE and PALIT (50), who studied nearly thirty liquids. These included aromatic and aliphatic hydrocarbons, alcohols, ketones and halogenated compounds. The monomer itself has a low transfer constant ($C = 2.6 \times 10^{-5}$ at 60°). Most other liquids are in the range from 1×10^{-4} to 1×10^{-3} but s-butyl alcohol has a transfer constant of 9.7×10^{-3}. Ease of transfer to amines was mentioned in a previous section, and Imoto (76) gives 1.04 for N,N-dimethylaniline. The absolute value of all these numbers must be accepted with some reservation because the basic equations relating intrinsic viscosity to number average molecular weight have undergone revision. Heterogeneity of the polymerization media may cause uncertainties also. It is clear that the relative values do not parallel those for other monomers (e. g. styrene). Bamford, Jenkins and Johnston (14) show that this difference is related to the powerful electron accepting character of the polyacrylonitrile radical.

Telomers have been prepared by polymerizing acrylonitrile in the presence of agents like CCl_4. Fox and Field (58) have reviewed this whole subject recently.

5. Polymerization in Aqueous Suspension

Aqueous systems have been studied by a very large number of investigators. Economy, safety, convenience and quality of product have combined to make this the method of choice for commercial production of copolymers. The industrial importance of such end products as elastomers and acrylic fibers has been a special incentive to related fundamental studies. Furthermore, the relatively high solubility of acrylonitrile monomer in water coupled with insolubility of the polymer make it a convenient test monomer for studies of initiation by redox systems (6, 25, 102). Large numbers of homogeneous chemical initiators and some heterogeneous initiators have been studied as well as initiation by photochemical means, by ultrasonics and by ionizing radiation. It will not be possible here to review the enormous world literature. Several publications (1, 92, 117) refer in some detail to the older papers, and we shall restrict our comments to recent interpretations that have received support from several quarters.

If radicals are produced in an aqueous solution of acrylonitrile there may be a short induction period during which air or impurities are being consumed. Smeltz and Dyer (120) have shown that polymeric peroxides are formed but that none of the ordinary high polymer is produced during this period. Onset of polymerization is evident from the development of a haze, followed rapidly by the precipitation of white polymer. The slurry increases in thickness, and by the time polymerization is complete it may be quite stiff, looking much like cottage cheese. The polymer generally filters readily but holds about four times its weight of water in the filter cake. This sequence of events is typical, but at low electrolyte concentration and at slow rates of stirring there may be formed a relatively stable suspension that passes through ordinary filter paper.

Further knowledge of the events taking place is gained by microscopical study. Rapid motion of individual particles makes it difficult to observe the changes directly, but samples may be withdrawn periodically for examination by the usual techniques of light and electron microscopy. Fig. 3 and 4 show some of the stages in typical experiments.

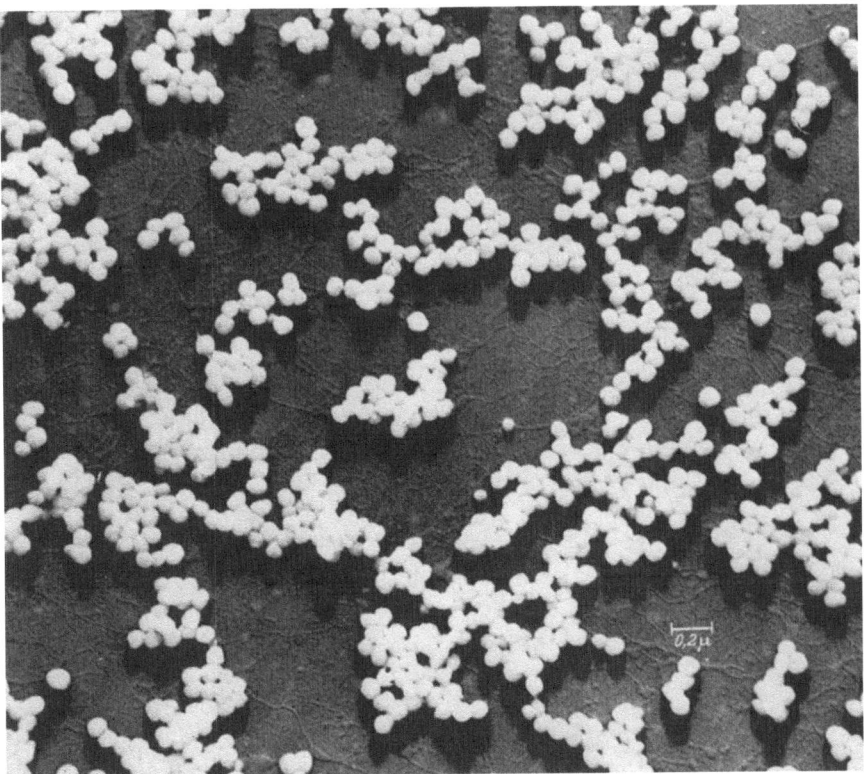

Fig. 3. Polyacrylonitrile prepared in aqueous emulsion. Electron micrograph, negative print, shadowed with uranium at a 30° angle. M. C. Botty and F. G. Rowe, American Cyanamid Company

More quantitative microscopical studies are reported by Thomas, Gleason and Mino (128) and by Nagao and Uchida (103). During the early stages of polymerization, particles separate as dense, slightly flattened spheroids occasionally joined by filaments or by partial fusion. The particles are often surprisingly uniform in size. They are known to have a high surface area, but details of their structure have not been disclosed. By the time the first observations have been made (about 1% conversion) the particles may have reached a diameter of 200 to 500 Å. From this point they grow until at high conversion the average diameter may be 1500 to 2500 Å. Few new particles, if any, are formed during this growth period, and in many experiments there is relatively little agglomeration. Particles are thought to be stabilized by a negative

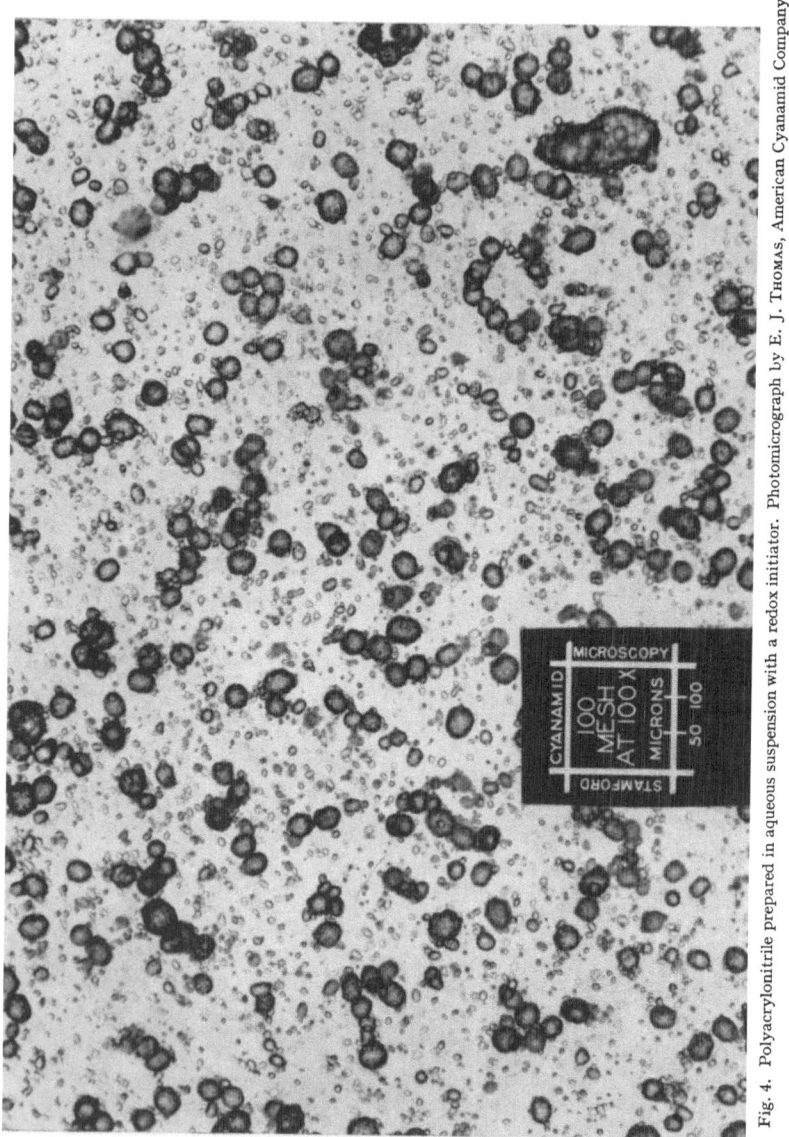

Fig. 4. Polyacrylonitrile prepared in aqueous suspension with a redox initiator. Photomicrograph by E. J. Thomas, American Cyanamid Company

charge and by hydration. Thus the number of particles per unit volume is often fairly constant during an experiment. This number nearly always lies between about 10^{12} and 10^{14} per cc., which is in the lower end of the typical range for emulsion polymerization in general. New particles appear if a second quantity of initiator is injected near the start of a run. The extent of aggregation is increased by addition of salts, or organic liquids, by agitation and by increased temperature. Aggregation is minimized by addition of surface active agents, and the number of particles per cc is increased by such additives. Recent experiments by YUGUCHI and WATANABE (137) show that agitation reduces the polymerization rate without affecting the induction period or the average degree of polymerization. As more sodium lauryl sulfate is added the effect on rate becomes less. Under conditions where the suspension aggregates without agitation the rate constants for polymerization could not be determined (49).

The general resemblance of these systems to typical emulsion polymerization has been stressed by several investigators. Thus the number of particles per cc can be varied by altering the concentration of initiator or dispersing agent, but the rate per particle is reasonably constant so long as aggregation is avoided. Dependence of polymerization rate on soap concentration has been established empirically in several cases.

PALIT and GUHA (110) drew further attention to the connection between polymerization rate and the colloidal nature of the precipitating polymer. They found that as the amount of redox initiator increased the polymerization rate first increased, then decreased and finally increased again. These regions corresponded to a fine sol, a milky dispersion and a coarse precipitate. Generally the rate of polymerization ran parallel to the amount of electrolyte required to precipitate the colloid.

Under suitable conditions a reasonably stable latex may be formed (1). Although many of the factors relating to polymerization rate, molecular weight and particle size in emulsion have been studied [reference (134) is an example] the literature on homopolymers is by no means as extensive as that pertaining to copolymers. Attempts to prepare bead homopolymers have not been very successful.

As is commonly the case, the deepest insight into the reaction mechanisms has been gained from kinetic studies. All of the usual techniques have been used, but that of DAINTON and SEAMAN (47) deserves special notice. These authors devised a dilatometer particularly suitable for the necessary deaerating, mixing, filling, stirring, polymerizing and cleaning steps. The percent polymerization vs. time curves often show a brief acceleration, followed by a nearly constant rate period and finally a region of reduced rate. The constant rate period may extend from a few-tenths percent up to 10—20% or even higher conversion depending on the initiator system and other factors. Under some conditions the acceleration period may be undetectable (47). In the case of photo-initiation or initiation by ionizing radiation there is a pronounced after-effect. When illumination ceases the rate drops rapidly at first, then more slowly (48). If the after-effect is observed at a temperature higher than the original reaction (50° vs. 25°) there is a burst of activity followed by a decay period (49), similar to that occurring in bulk polymerization but much less marked.

The most significant kinetic analysis is that of DAINTON and coworkers (46—49). Under conditions where termination by metal ions was not important, R_p varied as $[M]^2$ at low monomer concentrations and as $[M]$ at high concentrations. The critical monomer concentration for transition from one region to another varied with the rate of initiation, first decreasing and then increasing as the rate of initiation increased. Data in the literature giving exponents of $[M]$ of either one and two are apparently reconciled in this way.

Because of the heterogeneous nature of the polymerization it is clear that ordinary homogeneous kinetics cannot apply. Despite this, a mechanism leading to a stationary state can be developed (46) based on the observation mentioned above that a constant number of growing particles is formed near the start (128). The steady rate of polymerization is then the sum of the rate in the aqueous phase and that of monomer adsorbed on the polymer particles. When polymerization is chiefly in the aqueous phase (i. e., at high $[M]$), then the rate R_p depends on $[M] [I]^{1/2}$. When the particles are the principal locus (i. e., at low $[M]$), then R_p will vary as $[M]^2 [I]^X$, where $0.2 < X < 0.9$. If the rate of initiation is very high then R_p will depend on $[M]^2$ because of primary radical termination. Analysis of the after-effect suggests that radicals on the particles disappear slowly by mutual interaction and have a half-life at 25° of about 5 min. Radicals in solution have a corresponding half-life of about one second. Agitation facilitates the interaction of particles and so diminishes the after-effect. At the start of reaction radicals are produced in solution, where they grow until terminated in pairs. These unstable polymer molecules aggregate until they form particles having a size and charge such as to make coalescence of particles unlikely. Radicals may enter these particles, however, and cause polymerization of monomer adsorbed on the particles. Unimolecular termination has been suggested by some studies (132).

The general question of polymerization on or in the particle vs. polymerization in the true aqueous phase has been discussed for a number of years. Evidence other than that cited above suggests that both sites may be important depending on conditions. Polymer chains might be initiated in the aqueous phase and then grow further after transport to the particles. Unreported work in this Laboratory suggests that polymerization in the two sites concurrently, may account for some results on molecular weight distribution and on copolymerization. MINO (100) has shown from his own experiments and those of others that the partition of acrylonitrile between the two phases must be taken into account in copolymerization with styrene.

Preferential wetting of polymer by monomer was demonstrated qualitatively some years ago (60). Monomer adsorption has been confirmed by MILLER [cited in reference (128)], but quantitative information on the extent of adsorption is lacking. DAINTON makes the reasonable assumption that the concentration of monomer on the particle is proportional to that in solution.

By means of the sector technique DAINTON and EATON (49) determined the absolute values for the velocity coefficients applicable to the aqueous phase. Table 4 lists their results, which may be compared with those of Table 1 for polymerization in DMF. It appears that in aqueous media polyacrylonitrile radicals propagate very rapidly (comparable to polyacrylamide) and terminate quite rapidly. It will be noted that the activation energy for termination is higher than that for propagation. The frequency factor is higher for termination than is usually the case in polymerization reactions, but the frequency factor for propagation is normal. The large differences between coefficients in water and in DMF has not been explained adequately [see reference (46), p. 228]. Unfortunately there are no data for other homogeneous solvents. The work of BAWN (24) in heterogeneous systems certainly suggests that k_2 is highly sensitive to the environment. On the other hand in this event one would expect copolymer compositions to be more sensitive to the choice of reaction medium than they seem to be (133).

THOMAS and WEBB (129), using an emulsion polymerization model, calculated k_2 from R_p and the number of particles. Their value at room temperature was 2×10^4 l m^{-1}l^{-1}, in good agreement, perhaps fortuitously, with DAINTON and EATON (49). This treatment assumes that the monomer : water ratio at the particle is the same as in the continuous phase. If allowance were made for adsorption of monomer this value of k_2 would fall to approach more closely the level observed in DMF.

Table 6. *Rate Constants in Aqueous Polymerization (49) (Values Applicable to the Aqueous Phase)*

Temp. (°C)	Propagation $k_2 (\times 10^{-4})$	$k_4 (\times 10^{-9})$
15	2.3	2.8
25	2.8	3.7
30	3.25	4.4

Arrhenius Parameters

$$k_2 = 3.7 \times 10^7 \ e^{-4,100/RT}$$
$$k_4 = 3.3 \times 10^{13} \ e^{-5,400/RT}$$

Polymerization of acrylonitrile in water by means of ionizing radiation has been studied by a number of people over the past 15 years. The literature was reviewed in 1955 by CHAPIRO and coworkers (38). More recent reports are those of Bensasson and Prevot-Bernas (28), Chapiro and Sebban-Danon (39) and Arthur, Demint and Pittman (4) among others. Initiation is by H atoms and OH radicals derived from the water. It was thought originally that the polymerization kinetics demonstrated a non-uniform spatial distribution of initiating species (42). The present interpretation is that radiation does indeed yield primary radicals in tracks and clusters, but that the polymerization kinetics do not reflect these effects and can be explained adequately on the basis of the general picture given above (28, 47). The intensity exponent is in the neighborhood of 0.8 vs. 0.55 for monomer in DMF (28). Polymerization rate depends in a complex way on the water content of the system (4, 28, 42). Polymer made at high intensities in water may undergo cross-linking and degradation (4).

Continuous methods of polymerization have been studied extensively in industrial laboratories (96, 116). Advantages include greater production rates and much closer control of molecular weight. Although detailed knowledge of the mechanism is still lacking, Mintzer and Coman (101) have supplied data on the influence of such variables as initiator, temperature, residence time and pH. Conditions described by Mallison (96) lead to slurries with much different physical characteristics from those encountered in ordinary batch operation. With proper agitation he obtained fluid slurries with 40% polymer or more, and these slurries could be centrifuged to a water content as low as 25% of the cake weight. These features are correlated with the formation of dense, rounded particles often 10 to 50 microns across rather than the loose aggregates found in most batch processes. Rapid reaction rates in such systems suggest that a major fraction of the polymer is forming on the polymer particles.

6. Anionic Polymerization

It has been known for many years that acrylonitrile polymerizes readily in the presence of strong bases, but the detailed mechanisms are not well understood. Some of the difficulties are those common to all ionic polymerization studies, while others are due to the heterogeneous conditions and side reactions that are sometimes encountered. In the last two or three years interest has increased considerably and substantial progress can be anticipated.

Acrylonitrile is subject to initiation by bases because the strongly electron withdrawing CN group creates an electron-deficient double bond. The resulting anion is stabilized by the CN group. One would expect that in a series of vinyl monomers, $CH_2=CHX$, the ease of anionic polymerization should vary as the electron accepting ability of X. Conversely, with a given monomer the overall polymerization rate might vary as the base strength of the initiator. Both of these surmises are true, at least in general, but are not yet supported by extensive quantitative data for the acrylonitrile case. The order of reactivity among common monomers in anionic polymerization appears to be $CH_2=C(CN)_2 > CH_2=CHNO_2$

$$> CH_2=CHCN > CH_2=C\begin{matrix} CH_3 \\ \diagdown \\ CN \end{matrix} > CH_2=C\begin{matrix} CH_3 \\ \diagdown \\ COOCH_3 \end{matrix} > CH_2=CHC_6H_5.$$

Correlations of initiator base strength with polymerizability of several monomers are found in the work of HIGGINSON and WOODING (70). All of their bases were strong enough to polymerize acrylonitrile and so do not define a range.

FURUKAWA and coworkers (63) have shown this dual relationship in the form of a chart. They use the e-value in copolymerization as a measure of the electron deficient character of the double bond and the electronegativity of the metal as a measure of base strength of the organometallic initiator. Although these guides are useful they are only approximate. Temperature, solvent, side reactions and nature of the organic part of the initiator will be critical in each case.

Table 7 lists a number of agents that appear to initiate true anionic polymerization of acrylonitrile. Many more agents have been proposed,

Table 7. *Examples of Presumed Anionic Initiators*

Initiator Type	Example	Reference
Alkoxides	$NaOC_2H_5$	*34, 141*
Amides	KNH_2	*34, 53*
Alkyls or aryls	LiC_4H_9	*34, 78*
Salts	$CaZn(C_2H_5)_4$	*82*
Alfin catalyst	NaC_5H_{11}, etc.	*57, 78*
Grignard reagents	C_2H_5MgBr	*125, 26*
Phosphines	$P(C_2H_5)_3$	*72*
Ketyls	$Na(C_5H_5)_2CO$	*140, 81*
Sodiomalonic ester	$NaCH(COOEt)_2$	*45*
Ionizing radiation	Species unknown	*95*
Modified Ziegler	Cr acetylacetonate plus $ZnEt_2$	*106*
Quaternary ammonium hydroxides	$C_6H_5CH_2\overset{+}{N}(CH_3)_3O\overset{-}{H}$	*119*
Phosphites	$(EtO)_3P$	*51*
Hydrocarbon salts	$NaC\equiv CH$	*70*

but most of these are poorly understood or else appear on further study to be free radical sources. It is quite difficult to demonstrate that the growing chain end is a carbanion, particularly since classical methods like copolymerization studies are now known to be ambiguous. Reliance is usually placed on the chemical nature of the initiator, the low temperature range that is suitable, inhibition by water and CO_2 but not by hydroquinone and the like and sometimes on the colors one observes. In a typical anionic polymerization of acrylonitrile the following characteristics may appear:

Polymerization may be rapid or even explosive at room temperature and may be fairly rapid at dry ice temperature.

The polymer may be white (if made at low temperature) but is often yellow or yellow-brown.

Molecular weights are often very low (1000 or less) but may be in the 50,000 to 1,000,000 or even higher range under special conditions.

Polymer is sometimes wholly or partly soluble in acetone and other polar liquids that are not solvents for free-radical polymer.

Initiator fragments are sometimes incorporated.

The spectrum is often normal but may show evidence of alpha cyanoethylation and other structural differences, particularly in the case of acetone-soluble polymers.

One of the first detailed studies on these systems was that of BEAMAN (26), who showed that methacrylonitrile polymerizes by an anionic chain mechanism when treated with various bases, including Na in liquid ammonia at $-75°$ C. He noted also that low molecular weight polymers are obtained from reaction of acrylonitrile with butylmagnesium bromide. FOSTER (56) extended the liquid ammonia method to copolymerization studies in which acrylonitrile was combined with styrene, with methyl methacrylate and with vinyl butyl sulfone. Satisfactory data were obtained only with the sulfone, in which case there was some tendency for alternation.

HIGGINSON and WOODING (70) showed that acrylonitrile polymerizes readily in the presence of a large number of bases both in ether at $20°$ and in NH_3 at $-33.5°$. These bases covered a pK range of 17 to 36 and the work demonstrated that acrylonitrile is polymerized by bases too weak to initiate methyl methacrylate or styrene. The polyacrylonitriles were soluble in NH_3 and sparingly soluble in acetone and in pyridine. Although the authors thought that solubility might be associated with low molecular weight, it seems far more likely now that this was due to structural differences such as cyanoethylation. HIGGINSON and WOODING thought that the color of their polymers might be a result of conjugated structures arising from polymerization through the nitrile. This 1,4 polymerization has been proposed in other systems for methacrylonitrile (64) and for acrylonitrile (40, 86) but its relationship, if any, to color is not established.

Kinetic studies have been infrequent but CUNDALL (45) reported recently on the system acrylonitrile-DMF-sodium diethyl malonate. Designating the diethyl malonate anion as C^- and the DMF solvent as XH he showed that the rate expression is:

$$-d[M]/dt = k[M]^2 [C^-].$$

The data are consistent with the scheme:

$$C^- + M \ \rightarrow \ CM^- \tag{1}$$

$$CM^- + M \ \rightarrow \ CM_2^- \ \text{etc.} \tag{2}$$

$$CM_n^- + M \ \rightarrow \ CM_n + M^- \tag{3}$$

$$CM_n^- + XH \rightarrow \ CM_n + X \tag{4}$$

The average degree of polymerization is limited to about 20 by the transfer reaction (equation 3). There is about one double bond per polymer molecule, and infrared evidence suggests some incorporation of malonic ester fragments. Only the addition compound was formed in alcohol and tetrahydrofuran gave an inactive complex with the malonate. In general results were parallel to those on styrene in the pioneering kinetic study by HIGGINSON and WOODING (70). In the styrene work KNH_2 was the initiator and NH_3 was the solvent. Then termination was largely by transfer to solvent.

Recently, some detailed information has been provided on several classes of initiators in heterogeneous systems. One of these is a survey of ZILKHA, FEIT and FRANKEL (141) of sodium alkoxides. Acrylonitrile in light petroleum at $-15°$ (two liquid phases) was treated with sodium alkoxide solutions in the parent alcohol. The yield of polymer was higher and the induction period was shorter in the case of the more strongly basic alkoxides, i. e., those derived from less acidic alcohols like i-propyl and n-butyl. Similarly, the addition of i-propyl alcohol to methanolic sodium methoxide increased the activity of the methoxide. Larger amounts of catalyst gave greater yields until the total amount of alcohol present became excessive and the yield dropped.

Butyl lithium as an initiator was reported in another paper from the same laboratory (59). Working again in light petroleum at $-15°$, they found that a certain threshold concentration of C_4H_9Li was needed, but that this was less than for methyl methacrylate. Larger amounts of C_4H_9Li gave more polymer, but the molecular weight did not change much. Under these conditions the intrinsic viscosity was about 0.4 (molecular weight about 20,000). A higher yield was obtained at this temperature than at $-50°$, at $10°$ or at $35°$, but molecular weight was related inversely to temperature. Quantitatively this corresponded to a difference of 4 kcal per mole between the activation energies for propagation and termination. Use of tetrahydrofuran or ether as solvent led to lower yields than were obtained in light petroleum. These authors were able to prepare what appears to have been a block copolymer by adding acrylonitrile to a "living" styrene polymer initiated by C_4H_9Li.

In a third report ZILKHA and associates (140) regarded metal ketyls as species having both a free radical and an anion in the same molecule.

These two initiating sites might be considered to compete for monomer, but it appears that initiation is at the anionic site. In this study sodium benzophenone was used as initiator in tetrahydrofuran at 0°. Anionic polymerization was suggested by infrared evidence, by the rapid polymerization and by obtaining nearly pure polyacrylonitrile in attempted copolymerization with styrene. INOUE, TSURUTA and FURUKAWA (81) show that initiation is sensitive to specific reaction conditions. In their experiments sodium and potassium benzophenone gave polymer, but the lithium compound was ineffective.

Many organo-metallic compounds of the second, third and fourth groups of the periodic table have been suggested as initiators. In some cases these have been thought to be anionic but the evidence is often not clear and in certain cases it points to a free-radical mechanism. Tetraethyl lead, for example, requires light at room temperature, suggesting a radical process (97).

Conventional Ziegler catalysts are not suitable for use with acrylonitrile because, among other reasons, the monomer reacts with the catalyst or forms complexes with it. Recently, modified catalysts have been developed in NATTA's laboratory (106), using such combinations as:

Chromium acetylacetone plus dibutyl zinc,

Triethoxyvanadyl plus diethyl zinc,

Chromium acetylacetone plus triethyl aluminum.

Reaction takes place at moderate temperatures (40 to 80°) in a hydrocarbon or even in DMF. The polymer is described as a white or slightly colored product containing chain segments that show a syndiotactic configuration of monomer units. It is presumed by analogy with other Ziegler-Natta catalysts that the mechanism here is anionic.

Initiation by Lewis acids, although superficially cationic, must be dismissed as free-radical. If acrylonitrile dissolved in dimethylformamide is heated with a little BF_3, fairly rapid polymerization occurs. Other amide type solvents containing the N-methyl group can be used, and other acidic substances such as H_2SO_4, $SnCl_4$, $AlCl_3$ and SO_3 can be substituted for BF_3. Although these are typical ionic catalysts, CAMPBELL (36) showed that a free-radical polymerization is involved. Evidence includes the facts that oxygen is essential and that rather large amounts of water may be present. Initiation is thought to involve intermediates like:

$$\cdot O_2CH_2\overset{+}{N}=\overset{\overset{\displaystyle -BF_3:O}{|}}{\underset{\underset{\displaystyle CH_3}{|}}{C}}-H + HO_2 \cdot$$

The trialkyl borons comprise a class of initiators that had been regarded as ionic (5, 62, 88). However, it seems more reasonable now to look on these compounds as free-radical initiators (55). If polar monomers like vinyl acetate, vinyl chloride, methyl methacrylate or acrylonitrile are treated in solution with triethyl boron at room temperature, polymers are obtained in reasonably good yield. In some cases, at least, the polymers seem to be more highly ordered than are those derived from conventional initiators, but with acrylonitrile the evidence for high order is not convincing. The initiating species appear to be partially oxidized compounds like R_2BOOR derived from small amounts of air or from additives such as metal oxides. FORDHAM and STURM (55) point out that all monomers polymerized by trialkyl borons are polymerized also by conventional free-radical initiators, but not all of these monomers are susceptible to conventional ionic initiators. Furthermore, the boron compounds are said to be effective in aqueous systems, where an ionic mechanism is unlikely. As a third piece of evidence, these authors present copolymerization data for acrylonitrile and for styrene, with methyl methacrylate as the comonomer in each case. Copolymer composition for initiation by $B(C_4H_5)_3$ and by $B(C_2H_9)_3$ is in line with that of copolymers made with peroxides.

In contrast to the boranes, the tertiary phosphines appear to function as true anionic initiators (43, 72). Evidence for this viewpoint includes the following observations:

1. Monomers with electron withdrawing groups are polymerized by phosphines (acrylic and methacrylic esters, acrolein, methyl vinyl ketone, β-nitrostyrene, nitroethylene, acrylonitrile), whereas the phosphines have no effect on monomers more susceptible to acid catalysts (isobutylene, styrene).

2. Polymerization is vigorous even at $-70°$.

3. Oxygen has no effect other than to oxidize the phosphine.

4. Water lowers the yields of polymer and leads to low molecular weights.

5. Stable 1:1 adducts are formed if the monomer has very strong withdrawing groups, e. g., $CH_2=C(CN)_2$.

Unfortunately there seems to be no supplementary evidence on copolymerization initiated by phosphines. In a typical experiment, a solution of acrylonitrile in petroleum ether is treated with a little triethyl phosphine. A greenish-brown product settles on the walls of the flask. It is soluble in acetone, furfural, methyl ethyl ketone and DMF. Polymer precipitated from acetone by toluene is of low molecular weight and contains perhaps 0.5% combined phosphorus. In similar experiments it is possible to obtain pale yellow polymers with number average molecular weights of around 30,000 and with about one atom of P per polymer chain. On the basis of these results HORNER and coworkers proposed the scheme (72):

Initiation $\quad CH_2=CHCN \leftrightarrow \overset{\oplus}{CH_2}-\overset{\ominus}{CHCN} \rightarrow R_3\overset{\oplus}{P}CH_2\overset{\ominus}{CHCN}$

Propagation $\quad R_3\overset{\oplus}{P}CH_2\overset{\ominus}{CHCN} \rightarrow R_3\overset{\oplus}{P}(CH_2CHCN)_n CH_2\overset{\ominus}{CHCN}$

Termination $R_3\overset{\oplus}{P}(CH_2CHCN)_n CH_2\overset{\ominus}{CHCN} + H_2O \rightarrow HO\overset{\ominus}{R_3}\overset{\oplus}{P}(CH_2CHCN)_n CH_2CH_2CN$

Presumably, the growing chain would have a cyclic conformation or in some other way would avoid excessive charge separation.

Heisenberg and Jurgeleit (69) prefer triethyl phosphine as the initiator but find other alkyl phosphine to serve also. Aromatic phosphines are less effective, and one assumes that the rate of polymerization is directly related to the base strength of the phosphine. These investigators find that high molecular weight, pure white polymers can be obtained by addition of dry $CaCl_2$ to the reaction mixture. The role of $CaCl_2$ is not simply to act as a drying agent. Possibly it removes impurities that would lead to discoloration.

Dickey and Coover (51) showed that alkyl phosphites are capable of initiating polymerization of monomers bearing electron acceptor groups. Polymerization is rapid at moderate temperatures and, in the examples given, the polyacrylonitrile is soluble in acetone.

The most recent and most puzzling example of anionic initiation is that which seems to take place when monomer is subjected to ionizing radiation at low temperatures. Acrylonitrile polymerizes rapidly at temperatures well below 0° (and even in the solid at −196°) when irradiated with X-rays or with electrons. Since it is now known that cationic polymerization may be induced by radiation, there has been speculation that similar anionic polymerization might be occurring with acrylonitrile (87, 121). This subject has been discussed recently by Magat (95) who quotes unpublished work of Bensasson and of Tabata and Hiroshi.

A pertinent question is whether the structure of base-initiated polymer corresponds to the idealized head-to-tail arrangement that seems characteristic of radical-initiated polymer. General features of the infrared spectrum are identical for the two types. However, unreported work in this laboratory (119) suggests that structures of the type

$$+CH_2C+_{\overline{n}} \quad \begin{matrix} CH_2CHCN \\ / \\ \\ \backslash \\ CN \end{matrix}$$

are often present in base-initiated polymer. This would account for solubility in acetone, lower softening points and the presence of a band at 1420 cm⁻¹ in the infrared spectrum. If polymer made with X-rays at low temperature has followed an anionic course, then the presence of the ketene imine structure (40) must be regarded as a point of difference. In this connection it may be noted that Bensasson (30) saw no infrared or X-ray distinction between polymers made with 1 m. e. v. electrons at 20° and at −196°. There is general agreement that initiator fragments are often more prominent in base-initiated polymer. Color of polymer made anionically is generally (though not always) at least pale yellow. This is often ascribed to the presence of condensed pyridine-type rings similar to those postulated in polymer that has been heated strongly, or to 1,4 polymerization (70).

Until recently there was little conclusive evidence regarding the degree of stereospecificity of base-initiated polymer. Generally speaking the products varied from completely amorphous to moderately well ordered in two dimensions (i. e., perpendicular to the chain axis) but poorly ordered in the chain direction. It was assumed that a tendency existed toward syndiotactic configuration but that the sequences were probably short. This situation existed despite the fact that much of the recent work on anionic polymerization had been inspired by the hope of making crystalline polymers and that claims to crystalline polyacrylonitrile had been put forth from time to time. Recent work in several laboratories, including that of NATTA (106), shows that polymer with varying degress of order can be prepared by presumably anionic processes. So far there is little evidence that one can make either completely random or completely ordered polyacrylonitrile in this way. Syndiotactic sequences of varying length and frequency appear to have been made, but X-ray and microscopical criteria for crystallinity are lacking.

References

1. ANONYM.: The chemistry of acrylonitrile, second edition. New York: Published by American Cyanamid Co. 1959.
2. ARCUS, C. L., and A. BOSE: Steric investigation of maleic-anhydride-styrene copolymer and poly(acrylonitrile). Chem. and Ind. 456 (1959).
3. ARNETT, L. M., and J. H. PETERSON: Vinyl polymerization with radioactive aliphatic azobisnitrile initiators. J. Am. Chem. Soc. 74, 2031—2033 (1952).
4. ARTHUR, J. C. jr., R. J. DEMINT and R. A. PITTMAN: High energy γ-irradiation of vinyl monomers. I. Radiation polymerization of acrylonitrile. J. Phys. Chem. 63, 1366—1368 (1959).
5. ASHIKARI, N.: Polymerization and copolymerization of olefins with trialkylboron catalysts. J. Polymer Sci. 28, 250—252 (1958).
6. BACON, R. G. R.: Reduction activation. A new polymerization technique. Trans. Faraday Soc. 42, 140—155 (1946).
7. BAMFORD, C. H., and A. D. JENKINS: Studies in polymerization. VI. Acrylonitrile; the behavior of free radicals in heterogeneous systems. Proc. Roy. Soc. A 216, 515—539 (1953).
8. — — Kinetics of vinyl polymerizations in heterogeneous conditions. J. Polymer Sci. 14, 511 (1954).
9. — — Studies in polymerization. IX. The occlusion of free radicals by polymers; physical factors determining the concentration and behavior of trapped radicals. Proc. Roy. Soc. A 228, 220—237 (1955).
10. — — D. J. E. INGRAM and M. C. R. SYMONS: Detection of free-radicals in polyacrylonitrile by paramagnetic resonance. Nature (London) 175, 894—895 (1955).
11. — — Kinetics of bulk polymerization of acrylonitrile. J. Polymer Sci. 20, 405—409 (1956).
12. — — R. JOHNSTON: Rate of initiation in vinyl polymerization. Nature (London) 177, 992—993 (1956).

13. BAMFORD, C. H., A. D. JENKINS and R. JOHNSTON: Studies in polymerization. XI. Reactions between polymer radicals and ferric chloride in non-aqueous media. Proc. Roy. Soc. A239, 214—229 (1957).

14. — — — Studies in polymerization. XII. Salt effects on the polymerization of acrylonitrile in non-aqueous solution. Proc. Roy. Soc. A241, 364—375 (1957).

15. — and E. F. T. WHITE: Preparation of block copolymers by the tertiary base method. Trans. Faraday Soc. 54, 268—277 (1958).

16. — A. D. JENKINS and R. JOHNSON: Kinetic effects of salts on vinyl polymerization. J. Polymer Sci. 29, 355—366 (1958).

17. — W. G. BARB, A. D. JENKINS and P. F. ONYON: The kinetics of vinyl polymerizations by radical mechanisms. New York: Academic Press 1958.

18. — and E. F. T. WHITE: Reactions of peroxides in N,N-dimethylformamide solution. Part I. The induced decomposition of benzoyl peroxide. J. Chem. Soc. 1860—1863 (1959).

19. — A. D. JENKINS, R. JOHNSTON and E. F. T. WHITE: The relation between the molecular weight and intrinsic viscosity of polyacrylonitrile. Trans. Faraday Soc. 55, 168—178 (1959).

20. — — — The termination reaction in the polymerization of acrylonitrile. Trans. Faraday Soc. 55, 179—184 (1959).

21. — — — Termination by primary radicals in vinyl polymerization. Trans. Faraday Soc. 55, 1451—1460 (1959).

22. — — and E. F. T. WHITE: The kinetics of block and graft copolymerization of acrylonitrile. J. Polymer Sci. 34, 271—285 (1959).

23. BARSON, C. A., J. C. BEVINGTON and D. E. EAVES: The reactivities of monomers towards the benzoyloxy radical. Trans. Faraday Soc. 54, 1678—1683 (1958).

24. BAWN, C. E. H., T. P. HOBIN and W. J. McGARRY: Les radicaux piégés et la polymérisation dans les systèmes hétérogènes. J. chim. phys. 56, 791—797 (1959).

25. BAXENDALE, J. H., M. G. EVANS and G. S. PARK: The mechanism and kinetics of the initiation of polymerization by systems containing hydrogen peroxide. Trans. Faraday Soc. 42, 155—169 (1946).

26. BEAMAN, R. G.: Anionic chain polymerization. J. Am. Chem. Soc. 70, 3115—3118 (1948).

27. BENGOUGH, W. I.: Velocity coefficients for the polymerization of acrylonitrile. J. Polymer Sci. 28, 475—476 (1958).

28. BENSASSON, R., et A. PRÉVOT-BERNAS: Polymérisation radiochimique de l'acrylonitrile en solution et distribution spatiale des centres actifs primaires. J. chim. phys. 53, 93—95 (1956).

29. — — Polymerization radiochimique de l'acrylonitrile. Quelque observations sur le post-effect. J. Polymer Sci. 30, 163—173 (1958).

30. — Discussion following a paper by BAWN and coworkers. J. chim. phys. 56, 797 (1959).

31. BERNSTEIN, I. A., E. C. FARMER, W. G. ROTHSCHILD and F. F. SPALDING: Studies on the gamma-radiation-induced polymerization of acrylonitrile. J. Chem. Phys. 21, 1303—1304 (1953).

32. BEVINGTON, J. C., and D. E. EAVES: Initiation in the polymerization of acrylonitrile. Trans. Faraday Soc. 55, 1777—1782 (1959).

33. BOOTH, C., and L. R. BEASON: Statistical analysis of fractionation data. J. Polymer Sci. 42, 93—112 (1960).

34. BULLITT, O. H., jr.: Basic catalysis of the polymerization of acrylonitrile. United States Patents 2,608,554 (to duPont), August 26, 1952.

35. BURNETT, G. M., and L. D. LOAN: Solvent participation in radical chain reactions. Part II. Rates of polymerization in benzene solution. Trans. Faraday Soc. **51**, 219—225 (1955).
36. CAMPBELL, T. W.: Polymerization of acrylonitrile by boron trifluoride in amide solvents. J. Polymer Sci. **28**, 87—96 (1958).
37. CARPENTER, A. S.: Recent developments in synthetic fibers. J. Soc. Dyers Colourists **65**, 469—478 (1949).
38. CHAPIRO, A., M. MAGAT, A. PRÉVOT-BERNAS and J. SEBBAN: Polymérisation radiochimique des monomères vinyliques. J. chim. phys. **52**, 689—698 (1955).
39. — and J. SEBBAN-DANON: Polymérisations radiochimiques aux intensites élevées. J. chim. phys. **54**, 776—779 (1957).
40. CHEN, C. S. H., N. COLTHUP, W. DEICHERT and R. L. WEBB: Formation of ketene-imine structures during polymerization of acrylonitrile with X-rays. Paper submitted to J. Polymer Sci.
41. CLELAND, R. L., and W. H. STOCKMAYER: An intrinsic viscosity-molecular weight relation for polyacrylonitrile. J. Polymer Sci. **17**, 473—477 (1955).
42. COLLINSON, E., and F. S. DAINTON: The γ-ray and X-ray induced polymerization of aqueous solutions of acrylonitrile. Disc. Faraday Soc. 212—226 (1952).
43. COOVER, H. W., jr., and J. B. DICKEY: Polymerization of monomeric vinyl compounds. United States Patent 2,675,372 (to Eastman Kodak Co.).
44. COYNER, E. C., and W. S. HILLMAN: The thermal dimerization of acrylonitrile. J. Am. Chem. Soc. **71**, 324—326 (1949).
45. CUNDALL, R. B.: Sodiomalonic ester as an anionic catalyst. Proc. Chem. Soc. **1958**, 170—171.
46. DAINTON, F. S., and (in part) P. H. SEAMAN, D. G. L. JAMES and R. S. EATON: The polymerization of acrylonitrile in aqueous solution. J. Polymer Sci. **34**, 209—228 (1959).
47. — — The polymerization of acrylonitrile in aqueous solution. Part I. The reaction catalyzed by FENTON's Reagent at 25° C. J. Polymer Sci. **39**, 279 — 297 (1959).
48. DAINTON, F. S., and D. G. L. JAMES: The polymerization of acrylonitrile in aqueous solution. Part II. The reaction photosensitized by Fe^{3+}, $Fe^{3+}OH^-$, Fe^{2+} and I^- ions. J. Polymer Sci. **39**, 299—312 (1959).
49. — and R. S. EATON: The polymerization of acrylonitrile in aqueous solution. Part III. The ferric ion-photosensitized reaction at 15, 25, 30 and 50° C. J. Polymer Sci. **39**, 313—320 (1959).
50. DAS, S. K., S. R. CHATTERJEE and S. R. PALIT: Studies in chain transfer. V. Acrylonitrile. Proc. Roy. Soc. **227 A**, 252—258 (1955).
51. DICKEY, J. B., and H. W. COOVER jr.: Polymerization with alkyl phosphite catalysts. United States Patent 2,652,393 (to Eastman Kodak Co.), September 15, 1953.
52. DURUP, J., and M. MAGAT: Sur la cinétique de polymérisation en milieu précipitant. J. Polymer Sci. **18**, 586—588 (1955).
53. EVANS, M. G., W. C. E. HIGGINSON and N. S. WOODING: The mechanism of polymerization reactions in liquid ammonia. Rec. trav. chim. **68**, 1069—1078 (1949).
54. FLORY, P. J.: Principles of polymer chemistry. Ithaca: Cornell University Press 1953.
55. FORDHAM, J. W. L., and C. L. STURM: Mechanism of trialkylboron initiated polymerization. J. Polymer Sci. **33**, 503—504 (1958).
56. FOSTER, F. C.: Carbanionic copolymerization studies. J. Am. Chem. Soc. **74**, 2299—2302 (1952).

57. Foster, W. E.: Anionic polymerization process. United States Patent 2,841,574 (to Ethyl Corporation) July 1, 1958.
58. Fox, R. B., and D. E. Field: Telomerization. A review of the literature. U.S. Naval Research Laboratory Report 5190, November 19, 1958.
59. Frankel, M., A. Ottolenghi, M. Albeck and A. Zilkha: Anionic polymerization of vinyl monomers by butyl lithium. J. Chem. Soc. 3858—3864 (1959).
60. Fryling, C. S., and E. W. Harrington: Emulsion Polymerization. Ind. Eng. Chem. 36, 114—117 (1944).
61. Fueno, T., T. Tsuruta and J. Furukawa: Molecular orbital considerations on the reactivities of vinyl compounds. II. Ionic polymerizability. J. Polymer Sci. 40, 499—510 (1959).
62. Furukawa, J., T. Tsuruta and S. Inoue: Triethylboron as an initiator for vinyl polymerization. J. Polymer Sci. 26, 234—236 (1957).
63. — — Catalytic reactivity and stereospecificity of organometallic compounds in olefin polymerization. J. Polymer Sci. 36, 275—286 (1959).
64. Grassie, N., and I. C. McNeill: Thermal degradation of polymethacrylonitrile. IV. Formation and decomposition of ketene-imine structures. J. Polymer Sci. 33, 171—182 (1958).
65. Grobe, V., and E. Spode: The photopolymerization if acrylonitrile in magnesium perchlorate solution. Naturwissenschaften 44, 560—561 (1957).
66. Ham, G. E.: Polymerization of acrylics. Textile Res. J. 24, 597—614 (1954).
67. — Chain transfer in acrylonitrile polymerization. J. Polymer Sci. 21, 337—340 (1956).
68. — Helical stereospecific polymerizations. J. Polymer Sci. 40, 569—571 (1959).
69. Heisenberg, E., and W. Jurgeleit: Process for polymerization of vinyl compounds in the presence of a tertiary phosphine. United States Patent 2,921,055 (to Vereinigte Glanzstoff-Fabriken AG.) January 12, 1960.
70. Higginson, W. C. S., and N. S. Wooding: Anionic polymerization. J. Chem. Soc. 1952, 760—774, 774—779.
71. Holland, V. F.: Crystalline morphology of polyacrylonitrile. J. Polymer Sci. 53, 572—574 (1960).
72. Horner, L., W. Jurgeleit and K. Klüpfel: Tertiary phosphines. III. Inception of anionotropic polymerization in olefins. Ann. 591, 108—117 (1955).
73. Howard, W. H., A. T. Watson and T. W. Dewitt: The glass transition temperature of polyacrylonitrile. Paper presented at the International Symposium on Macromolecules. Wiesbaden 1959.
74. Hunyar, A., and V. Grobe: On the polymerization of acrylonitrile in inorganic salt solutions. Faserforsch. u. Textiltech. 6, 548—553 (1955).
75. Imoto, M., and K. Takemoto: Vinyl polymerization. VI. Polymerization of acrylonitrile in the presence of substituted benzoyl peroxides and dimethyl aniline. J. Polymer Sci. 18, 377—387 (1955).
76. — T. Otsu, T. Ota, H. Takatsugi and M. Matsuda: Vinyl polymerization. XVI. Effect of dimethylaniline on rate of polymerization of vinyl monomers initiated by 2,2'-azobisisobutyronitrile. J. Polymer Sci. 22, 137—147 (1956).
77. Imoto, M., and H. Takatsugi: Vinyl polymerization. XVIII. Kinetics of the polymerization of acrylonitrile in benzene. Makromol. Chem. 23, 119—127 (1957); Kobunshi Kagaku 15, 65—70 (1958).
78. — M. Kinoshita: Polymerization of vinyl monomers with alfin catalysts. J. Chem. Soc. Japan, Ind. Chem. Sect. 61, 452—454 (1958).
79. — Vinyl polymerization. XXX. Heterogeneous polymerization rate of acrylonitrile in mixed solvent of benzene and dimethylformamide. J. Polymer Sci. 31, 195—197 (1958).

80. INGRAM, D. J. E., M. C. R. SYMONS and M. G. TOWNSEND: Electron resonance studies of occluded polymer radicals. Trans. Faraday Soc. 54, 409—415 (1958).

81. INOUE, S., T. TSURUTA and J. FURUKAWA: Metal ketyl as initiator for vinyl polymerization. Makromol. Chem. 36, 77—80 (1959).

82. — — — Catalytic activity of organometallic compounds for olefin polymerization. Makromol. Chem. 32, 97—111 (1959).

83. JENKINS, A. D.: Transfer to solvent and retardation in vinyl polymerization. Trans. Faraday Soc. 54, 1885—1901 (1958).

84. — R. JOHNSTON: Chain transfer to a mixture of agents: The anomalous behavior of the acrylonitrile-dimethylformamide-water system. J. Polymer Sci. 39, 81—86 (1959).

85. KEAVNEY, J. J., and E. C. EBERLIN: The determination of glass transition temperatures by differential thermal analysis. J. Appl. Poly. Sci. 3, 47—53 (1960).

86. KERN, W., and H. FERNOW: Über die Polymerisation des Acrylnitrils und Polyacrylnitril. J. prakt. Chem. 160, 281—295 (1942).

87. KOBAYASHI, H.: Molecular weight dependence of intrinsic viscosity, diffusion constant, and second virial coefficient of polyacrylonitrile. J. Polymer Sci. 39, 369—388 (1959).

88. KOLESNIKOV, G. S., and L. S. FEDOROVA: Polymerization of acrylonitrile in the presence of tributyl boron. Izvestiya Akad. Nauk. Otdel. Khim. Nauk 236—237 (1957).

89. KONINGSBERGER, C., and G. SALOMON: Preparation and properties of rubber-like high polymers. I. Polymerization of dienes and vinyl compounds in bulk. J. Polymer Sci. 1, 200—216 (1946).

90. KRIGBAUM, W. R., and A. M. KOTLIAR: The molecular weight of polyacrylonitrile. J. Polymer Sci. 32, 323—341 (1958).

91. — and N. TOKITA: Melting point depression study of polyacrylonitrile. J. Polymer Sci. 43, 467—488 (1960).

92. LACZKOWSKI, M., M. KAUCZYNSKA-WOLFRAM and J. PLOSZAJSKI: Untersuchungen über die Polymerisationkinetik von Acrylnitril in heterogenen System. Faserforsch. u. Textiltechn. 8, 358—363 (1957).

93. MAGAT, M.: Sur la cinétique de polymérisation en milieu précipitant. J. Polymer Sci. 19, 583—585 (1956).

94. — Amorçage des polymérisations par les rayons γ. Special issue of the collection of Czechoslovak Chemical Communications 22, 141—152 (1957).

95. — Sur les polymérisations ioniques amorcées par des rayonnements ionisants. Makromol. Chem. 35, 159—173 (1960).

96. MALLISON, W. C.: Continuous process for the polymerization of acrylonitrile. United States Patent 2,777,832, (to American Cyanamid Company), January 15, 1957.

97. MARVEL, C. S., and R. G. WOOLFORD: Lead tetraethyl as initiator for polymerization reactions. J. Am. Chem. Soc. 80, 830—831 (1958).

98. MILLER, M. L., P. A. BUTTON, R. F. STAMM, L. RAPOPORT and E. H. GLEASON: Paper presented at meeting-in-miniature, New York Section of the American Chemical Society, March 16, 1956.

99. — Block and graft polymers. I. Graft polymers from acrylamide and acrylonitrile. Can. J. Chem. 36, 303—308 (1958).

100. MINO, G.: Copolymerization of styrene and acrylonitrile in aqueous dispersion. J. Polymer Sci. 22, 369—383 (1956).

101. MINTZER, J., and M. COMAN: Die kontinuierliche Polymerisation des Acryl-nitrils. Faserforsch. u. Textiltech. 9, 1—10 (1958).

102. MORGAN, L. B.: Transient molecular species occuring in persulphate oxidations. Trans. Faraday Soc. 42, 169—183 (1946).

103. NAGAO, H., and M. UCHIDA: Effect of added reagents on the aqueous poly-merization of acrylonitrile. Study with an electron microscope. J. Chem. Soc. Japan Ind. Chem. Sect. 61, 466—469 (1958).

104. NAKATSUKA, K.: On the polymerization of acrylonitrile in solution. Third report. On the kinetics of the polymerization reaction. Chemistry High Poly-mers, Japan 15, (1), 43—48 (1958).

105. — Polymerization reaction of acrylonitrile in solution. Fourth report. Retarda-tion of the polymerization reaction by the use of nitro derivatives of phenol. Chemistry High Polymers, Japan 16, 453—455 (1959).

106. NATTA, G., and G. DALL'ASTA: Process for the polymerization of acrylonitrile and polymers obtained thereby. French Patent 1,183,644 (to Montacatini) published July 9, 1959.

107. OKAMURA, S., K. KATAGIRI and Y. TAKEMOTO: Effect of surface-active agents in the polymerization of acrylonitrile. J. Chem. Soc. Japan, Ind. Chem. Sect. 61, 241—243 (1958).

108. ONYON, P. F.: Molecular weights in acrylonitrile polymerization. J. Polymer Sci. 22, 19—23 (1956).

109. — The polymerization of acrylonitrile in dimethylformamide. Trans. Faraday Soc. 52, 80—88 (1956).

110. PALIT, S. R., and T. GUHA: Some observations on the effect of the physical nature of the separating phase on the rate of heterogeneous polymerization. J. Polymer Sci. 34, 243—250 (1959).

111. PEEBLES, L. H. jr.: Branching in polyacrylonitrile. J. Am. Chem. Soc. 80, 5603—5607 (1958).

112. — Kinetics of polymerization of acrylonitrile in solution. Paper presented at the Division of Polymer Chemistry of the American Chemical Society in Boston on April 5—10, 1959.

113. PRÉVOT, A.: Rayons X-dosage du rayonnement X par réactions de polymérisa-tion. Compt. rend. 230, 288—290 (1950).

114. — and J. CABANNÉS: Polymérisation en phase homogène en hétérogène: acrylonitrile en solution dans la diméthylformamide. Compt. rend. 237, 1686 —1688 (1953).

115. — and J. SEBBAN-DANON: Sur certains caractères de la polymérization radio chimique de l'acrylonitrile. J. Chem. Phys. 53, 418—421 (1956).

116. SCHEIDERBAUER, R. A., and L. S. PITTS: Polymerization. United States Patent 2,748,106, May 29, 1956 to E. I. duPont de Nemours and Company.

117. SCHILDKNECHT, C. E.: Polymer processes. Chapter V. New York: Inter-science Publishers Inc. 1956.

118. SCHMIDT, W. G.: Process for making acrylonitrile polymer solution. United States Patent 2,923,694, (to Courtaulds Ltd.) February 2, 1960.

119. SCHULLER, W. H., and coworkers. Unreported work at American Cyanamid Company.

120. SMELTZ, K. C.. and E. DYER: The effect of oxygen on the polymerization of acrylonitrile. J. Am. Chem. Soc. 74, 623—628 (1952).

121. SOBUE, H., and Y. TABATA: Two different polymerization mechanism of acrylonitrile initiated by γ-irradiation. J. Polymer Sci. 43, 459—465 (1960).

122. SRINIVASAN, N. T., and M. SANTAPPA: Polymerization of acrylonitrile. Makromol. Chem. **26**, 80—91 (1958).

123. STÉFANI, R., M. CHEVRETON, J. TERRIER and C. EYRAUD: Propriétés méchaniques et structure des polymères d'acrylonitrile. Compt. rend. **248**, 2006 —2008 (1959).

124. TAKATA, T., and M. TANIYAMA: Synthesis of model substances for polyacrylonitrile and its copolymers. I. Synthesis of model substances for polyacrylonitrile-1. Chemistry High Polymers, Japan **16**, 693—698 (1959).

125. TERENT'EV, A. P., and A. N. YOST: Action of organic magnesium compounds on acrylonitrile. Vestnik Moskov. Univ. Ser. Fiz.-Mat. i Estestven. Nauk **5**, 41—42 (1950).

126. THOMAS, W. M., and J. J. PELLON: Kinetics of acrylonitrile polymerisation in bulk. J. Polymer Sci. **13**, 329—353 (1954).

127. — E. H. GLEASON and J. J. PELLON: Acrylonitrile polymerization in homogeneous solution. J. Polymer Sci. **17**, 275—290 (1955).

128. — — and G. MINO: Acrylonitrile polymerization in aqueous suspension. J. Polymer Sci. **24**, 43—56 (1957).

129. — and R. L. WEBB: Propagation rate in acrylonitrile polymerization. J. Polymer Sci. **25**, 124—125 (1957).

130. — A. M. THOMAS and W. G. DEICHERT: Microscopical study of heterogeneous polymerization. Paper presented at the International Symposium on Macromolecules. Wiesbaden 1959.

131. TONG, L. K. J., and W. O. KENYON: Heats of polymerization of some unsaturates. J. Am. Chem. Soc. **69**, 2245—2246 (1947).

132. TSURUTA, M., F. KOBAYASHI and K. TOMITA: On the rate of polymerization of acrylonitrile in aqueous medium. J. Chem. Soc., Japan Ind. Chem. Sect. **62**, 1620—1622 (1959).

133. UCHIDA, M., and H. NAGAO: Effect of emulsifier upon the compositions of emulsion copolymers of acrylonitrile and water-soluble monomers. Bull. Chem. Soc. Japan **30**, 311—314 (1957).

134. — — Studies on acrylonitrile polymers. XII. Emulsion polymerization of acrylonitrile. J. Chem. Soc., Japan Ind. Chem. Sect. **60**, 484—488 (1957).

135. ULBRICHT, J.: Zur Kinetik der homogenen Lösungpolymerisation von Acrylnitril. Faserforsch. u. Textiltech. **10**, 115—120 (1959). Part II. ibid. 166—172.

136. HOOD, J. P. VAN, and A. V. TOBOLSKY: Thermal Decomposition of 2,2'-azo-bis-isobutyronitrile. J. Am. Chem. Soc. **80**, 779—782 (1958).

137. YUGUCHI, S., and M. WATANABE: Studies on acrylic fiber. X. Effect of agitation on polymerization of acrylonitrile in aqueous and emulsion phase initiated by persulfate-triethanolamine redox system and persulfate alone. Chem. High Polymers **17**, 108—114 (1960). See also previous papers in this series.

138. ZAHN, H., and P. SCHÄFER: Zur Kenntnis der Oligomeren des Acrylnitrils. I. Synthesen. Ber. **92**, 736—744 (1959).

139. ZIEGLER, K., W. DEPARADE and H. KÜHLHORN: Zur Kenntnis des „dreiwertigen" Kohlenstoffs. XXIV.: Einige Versuche über Polymerisations-Erregung durch Radikale. Ann. **567**, 151—179 (1950).

140. ZILKHA, A., N. PADATSUR and M. FRANKEL: Metal ketyls as initiators of polymerization of vinyl monomers. Proc. Chem. Soc. **1959**, 364.

141. — B.-A. FEIT and M. FRANKEL: Alkoxides as initiators of anionic polymerization of vinyl monomers. Part I. Polymerization of acrylonitrile by use of sodium alkoxides. J. Chem. Soc. 928—933 (1959).

Fortschr. Hochpolym.-Forsch., Bd. 2, S. 442—464 (1961)

Recent Progress in Silicone Chemistry.
I. Hydrolysis of Reactive Silane Intermediates

By

M. M. SPRUNG

General Electric Research Laboratory, Schenectady, N. Y.

Contents

	Page
I. Introduction	442
II. Intermediate Hydrolysis Products from Difunctional Monomers	443
III. Intermediate Hydrolysis Products from Trifunctional Monomers	444
Cohydrolysis with Monofunctional Compounds	447
IV. Silsesquioxanes	448
V. Intermediate Hydrolysis Products from Tetrafunctional Monomers	450
Silicic Acids	452
VI. Silanols	453
A. Silanemonols	453
B. Silanediols	454
C. Polysiloxanediols	455
D. Silanetriols	454
E. Silanetetrol	455
F. Reactions of Silanols	455
VII. Mechanism of Solvolysis of Reactive Silanes	457
A. The Silicon-Hydrogen Bond	457
B. The Silicon-Halogen Bond	459
C. The Si–OR Bond	461
References	461

I. Introduction

Silicone polymer technology rests in practice upon the preparation of reactive substituted silanes from silicon metal and the subsequent conversion of these reactive substances, usually through stepwise hydrolysis and condensation reactions, into polysiloxanes. Thus the hydrolysis of these reactive intermediates is a fundamental process, the nature and implications of which have demanded increasing attention as organosilicon chemistry and technology have developed.

General schemes for chlorosilane hydrolysis have been quite thoroughly elaborated in the literature. Significant contributions to this subject were made fairly early (*26, 28, 54*).

Hydrolysis of a reactive silane derivative under the proper conditions first produces a silanol, but except in the case of a few that are exceptionally stable, this product is not isolated, since the acid or alkaline reagents that are normally present cause most silanols to condense rapidly to form siloxanes. The product generally isolated when a dialkyl- or diaryl-substituted silane is hydrolyzed is, of course, a mixture of cyclic polysiloxanes, $(R_2SiO)_n$, from which high polymers (often elastomeric) are obtained by catalytic rearrangement and polymerization reactions.

II. Intermediate Hydrolysis Products from Difunctional Monomers

The hydrolysis of a dichlorosilane is a stepwise reaction. The chlorosilanol, first formed, reacts with excess chlorosilane, and under the proper conditions a series of α, ω-dichloropolysiloxanes can be obtained. They are, however, not very stable in the usual hydrolysis medium, where the tendency is to form the stable cyclic polysiloxanes. The use of polar solvents facilitates their isolation, and this procedure was used by PATNODE and WILCOCK (53) to prepare the first five members of the series $Cl(SiMe_2O)_nSiMe_2Cl$ (n = 1–4). Later, SHAFFER and FLANIGEN (66) employed a similar procedure to isolate the analogous diethyl derivatives, $Cl(SiEt_2O)_nSiEt_2Cl$ (n = 1–3). The first three members of the diphenyl series, $Cl(SiØ_2O)_nSiØ_2Cl$, were described by BURKHARD some time ago (13).

SHAFFER and FLANIGEN (66), using conductometric titrations as a tool, obtained indirect evidence for the formation, in highly polar solvents, of a variety of intermediates. For example, titration endpoints at H_2O/Me_2SiCl_2 ratios of 0.53 and 0.73 were indicative of the predominance in the reaction medium of $Me_2SiClOSiMe_2Cl$ and $Cl(Me_2SiO)_{\sim3}SiMe_2Cl$; while diphenyldichlorosilane gave an end-point at 2.0, corresponding to diphenylsilanediol.

Stepwise hydrolysis is much more feasible in the case of dialkoxydialkyl- or dialkoxydiarylsilanes, for one reason because these may be hydrolyzed in the presence of only catalytic amounts of acidic or alkline reagents. The simplest examples of this class were recently described by LASOCKI (37) who hydrolyzed dimethyldimethoxysilane, using 0.75 mole of water per mole of $Me_2Si(OMe)_2$ and dilute alcoholic methanol as the solvent, to obtain members of the series, $MeO(SiMe_2O)_nMe$ (n = 2–10). Some cyclic poly(dimethylsiloxanes) were formed simultaneously. The ratio of linear to cyclic products varied with the proportion of water to methoxysilane and with the amount of base used as catalyst. Similar observations were made by MATSUI (39).

FLETCHER and HUNTER (17) also found that both linear and cyclic products were formed simultaneously on hydrolysis of diethoxydimethylsilane, $Me_2Si(OEt)_2$. The series, $EtO(SiMe_2O)_nEt$ ($n = 1-11$) was described, and procedures were worked out for preparing low or high polymers and for minimizing the amounts of cyclic siloxanes formed.

More recently, work in this general area has been extended by Okawara and his collaborators (45—52), who describe the partial hydrolysis of diisopropoxydimethylsilane to give iso-$PrO(Me_2SiO)_nPr$ ($n = 1-5$) and of di-$tert$-butoxydimethylsilane to give $tert$-$BuO(Me_2SiO)_nBu$ ($n = 1-5$). Both reactions can be carried out in one step by treating Me_2SiCl_2 with a mixture of the alcohol and water in benzene-pyridine solution. Okawara and Sakiyama (52) then hydrolyzed dichloroethyl-silane in the presence of monofunctional chain blocking groups and obtained the compounds $Me_3SiO(SiHEtO)_nSiMe_3$ ($n = 1-5$) and the single disiloxane, $HSiMe_2OSiHEtOSiMe_2H$.

Careful hydrolysis of the α, ω-dichloropolysiloxanes gives the corresponding α, ω-polysiloxanediols. This was first demonstrated for the phenyl-substituted compounds, $HO(Si\varnothing_2O)_nH$, by Kipping and Robison (35) and much more recently for the methyl-substituted compounds, $HO(SiMe_2O)_nH$ ($n = 2-5$) by Sokolov (70).

An interesting series of chain-blocked poly(acetoxymethyl-methyl-siloxanes) was obtained by Andrianov and Makarova (6) through partial hydrolysis of acetoxymethyldiacetoxymethylsilane in the presence of acetoxytrimethylsilane.

$$AcOCH_2Si(Me)(OAc)_2 + Me_3SiOAc + H_2O \rightarrow Me_3SiO\left[\begin{array}{c} Me \\ | \\ SiO \\ | \\ CH_2OAc \end{array}\right]_{1-4}SiMe_3$$

By suitable variations, the acetoxymethyl groups can be replaced by methoxymethyl or ethoxymethyl groups.

An alternate method for the preparation of α, ω-dichloropolysiloxanes has been developed by Sokolov and Andrianov (71). This consists in the self-condensation or inter-condensation of chloroalkoxysilanes in the presence of ferric chloride; e. g.,

$$2\ R_2SiCl(OEt) \xrightarrow{FeCl_3} EtCl + ClSiR_2OSiR_2Cl$$

$$ClSiR_2OSiR_2Cl + ClSiR'_2OEt \longrightarrow EtCl + ClSiR_2OSiR_2OSiR'_2Cl\ \ etc.$$

III. Intermediate Hydrolysis Products from Trifunctional Monomers

The trihalogeno-substituted silanes are very readily hydrolyzed, and the initially formed silanols condense with themselves and with halo-genosilanes with extreme ease. Consequently, few simple intermediate hydrolysis products of this class have been isolated or characterized. Linear halogen-substituted silanols and siloxanes are certainly first formed, and these then condense further to give cyclic products or branched open-chain structures. The reaction scheme is formally similar to that observed for difunctional silanes, but much more complex because of the number of paths that can be, and undoubtedly are, followed. A schematic diagram must include interactions between innumerable pairs of hypothetical intermediates. A partial reaction scheme may be visualized, as shown below.

$$RSiCl_3 + H_2O \rightarrow RSiCl_2OH \rightarrow RSiCl(OH)_2 \rightarrow RSi(OH)_3$$

$$RSiCl_3 + RSiCl_2(OH) \rightarrow RSiCl_2OSiCl_2R$$

$$RSiCl_3 + RSiCl(OH)_2 \rightarrow RSiCl_2OSiCl(OH)R$$

$$RSiCl_2OSiCl_2R + RSiCl_2OH \rightarrow RSiCl_2OSi(R)ClOSiCl_2R$$

$$RSiCl_2OSi(R)ClOSiCl_2R + RSiCl_2(OH) \rightarrow RSiCl_2OSi(R)OSiCl_2R$$
$$\qquad\qquad\qquad\qquad\qquad\qquad\qquad\qquad\qquad\underset{|}{OSiCl_2R}$$

$$RSiCl_2OSi(R)ClOSiCl_2R + RSiCl(OH)_2 \rightarrow$$

$$RSiCl_2OSi(R)ClOSi(R)ClOSi(R)Cl(OH)$$
$$\qquad\qquad\qquad\downarrow$$
$$\qquad\qquad(RSiClO)_4 \quad \text{(cyclic tetrasiloxane)}$$
$$\qquad\qquad\qquad\text{etc. etc.}$$

WEST (90) studied the hydrolysis of the simplest member of the series, trichlorosilane, and reported yields of 5—10% of the linear dimer, $HCl_2SiOSiHCl_2$, when the hydrolysis was carried out in ether at —78°. Methyltrichlorosilane gave only traces of distillable substances under these conditions.

SHAFFER and FLANIGEN (66) using a conductometric method, found evidence for the presence in solution of a number of the above intermediates in a system consisting of a trichlorosilane (methyl-, amyl-, phenyl-, or vinyltrichlorosilane), water, and a polar solvent. Only the symmetrical disiloxane, $C_6H_5SiCl_2OSiCl_2C_6H_5$, and the mixed disiloxane, $C_6H_5SiCl_2OSiCl_2CH=CH_2$, were actually isolated. However, with methyltrichlorosilane, conductometric end-points at H_2O/chlorosilane ratios of 1.02 and 1.5 presumably defined the cyclic species, $(MeClSiO)_x$ and the cross-linked species, $(MeSiO_{1.5})_x$. With phenyltrichlorosilane, end-points at 0.54 and 1.6 suggested the presence of $C_6H_5SiCl_2OSiCl_2C_6H_5$ and $(C_6H_5SiO_{1.5})_x$. Mixtures of phenyltrichlorosilane and dimethyldichlorosilane gave indications of the formation of $C_6H_5Si(OSiMe_2Cl)_3$, $C_6H_5SiCl_2$–$OSiMe_2Cl$, and $[C_6H_5Si(OSiMe_2Cl)_2]_2O$.

The use of tertiary amines as acid acceptors tends to stabilize some of the intermediate products and to simplify some of the problems of isolation. However, there are as yet few published reports on this phase of the problem.

The hydrolysis of trialkoxysilanes is somewhat easier to control, but even here the simplest members of the series are still so reactive that isolation of low molecular weight intermediates is difficult. Thus Kantor (33) found that methylsilanetriol could not be obtained from methyltrimethoxysilane even when the hydrolysis was carried out with pure water and in an all quartz apparatus.

Sprung and Guenther (74) treated methyltriomethoxysilane in benzene solution with three molar equivalents of water, and isolated only a water soluble, highly hydrated "poly(methylsiliconic acid)."

With lower ratios of water to methyltrimethoxysilane, both linear and cyclic products,

$$\mathrm{MeO} \begin{bmatrix} \mathrm{Me} \\ | \\ \mathrm{Si{-}O} \\ | \\ \mathrm{OMe} \end{bmatrix}_n \mathrm{Me} \quad \text{and} \quad \begin{bmatrix} \mathrm{Me} \\ | \\ {-}\mathrm{Si{-}O{-}} \\ | \\ \mathrm{OMe} \end{bmatrix}_n$$

were isolated. Tamborski and Post (84) had previously obtained the linear dimer, trimer and tetramer in a similar experiment. A variation of this technique, described by Tanaka, Tasaka, Okawara and Watase (85) is the partial methanolysis of methyltrichlororosilane, followed by neutralization and distillation.

Okawara (46) has discussed the use of infrared as a diagnostic tool in this relation.

Linear ethoxymethylsiloxanes were first described by Andrianov (4) and later by Fletcher and Hunter (18). Okawara, Minami and Oku (51) employed the technique of partial hydrolysis of dichloroethoxymethylsilane in the presence of pyridine to prepare the linear dimer and trimer,

$$\begin{array}{cc} \mathrm{Me} \ \ \mathrm{Me} \\ | \quad | \\ \mathrm{Et\,O\,Si\,O\,Si\,O\,Et} \\ | \quad | \\ \mathrm{OEt\ OEt} \end{array} \quad \text{and} \quad \begin{array}{c} \mathrm{Me} \ \ \mathrm{Me} \ \ \mathrm{Me} \\ | \quad | \quad | \\ \mathrm{Et\,O\,Si\,O\,Si\,O\,Si\,O\,Et} \ . \\ | \quad | \quad | \\ \mathrm{OEt\ OEt\ OEt} \end{array}$$

Sprung and Guenther (75) also obtained similar products under suitable conditions, but on varying the monomer-water ratio, were able to demonstrate the presence of more complex hydrolysis products in which cyclopolysiloxanes containing ethoxyl and hydroxyl groups are fused together or linked together by means of oxygen or siloxane units.

The problem of isolation of the more highly condensed structures becomes somewhat less severe when the alkyl-substitutuent on silicon is more bulky. Upon hydrolysis of ethyltriethoxysilane, Sprung and Guenther (76) were able to demonstrate the presence of multi-ring structures having ethoxyl and hydroxyl end groups, in considerably higher yield than with the methyl compound. With lower ratios of water to trifunctional silane, less highly cyclic products and ultimately the straight chain, ethoxy end-blocked poly(ethoxyethylsiloxanes) predominated. The latter were also obtained by OKAWARA (45) on treatment of dichloroethoxyethylsilane with appropriately small amounts of water.

The presence of bulky alkoxy groups also increases the stability of the monomer, but at the same time increases the difficulty of isolating intermediate products, since the removal of the alkoxy groups produces species that are more and more suspectible to condensation. OKAWARA and ISHIMARU (50) treated ethyltrichlorosilane with isopropyl alcohol and isolated only the alkoxy-blocked, linear dimer and trimer and the cyclic trimer, triethyltriisopropoxycyclotrisiloxane. An apparently unique product containing a single *n*-butoxymethylsiloxy group "trapped" in an otherwise completely condensed polycyclic structure was isolated in low yield by SPRUNG and GUENTHER on hydrolysis of methyl-tri-*n*-butoxysilane (77).

The presence of still larger alkyl or aryl groups on silicon may stabilize low molecular weight hydrolysis products against further attack. On hydrolysis of *n*-amyl- and phenyltriethoxysilanes, SPRUNG and GUENTHER (78) found evidence for the formation of cyclic polysiloxanes (in the cyclotetrasiloxane through cyclooctasiloxane range) with an average of 2—4 uncondensed silanol groups per molecule. Under somewhat different conditions, crystalline solids were isolated from phenyltriethoxysilane corresponding empirically to $8C_6H_5SiO_{1.5} \cdot 2EtOH$. Since the formation of the cyclotetrasiloxane ring is often favored through both steric and energetic considerations, a structure was suggested consisting of three fused cyclotetrasiloxane rings bearing hydroxy and ethoxy end groups,

$$
\begin{array}{cccc}
\varnothing & \varnothing & \varnothing & \varnothing \\
| & | & | & | \\
RO-Si-O-Si-O-Si-O-Si-OR \\
| & | & | & | \\
O & O & O & O \\
| & | & | & | \\
RO-Si-O-Si-O-Si-O-Si-OR \\
| & | & | & | \\
\varnothing & \varnothing & \varnothing & \varnothing
\end{array}
$$

$$R = H \text{ or } Et.$$

Cohydrolysis with Monofunctional Compounds. The effective functionality can be reduced to two (or less, if desired) by cohydrolysis of a trifunctional and a monofunctional silane. This method was used by ANDRIANOV, et al. (5) to prepare linear cohydrolysis products, according to the equation:

$$(X + 2)R_3SiCl + XArSiCl_3 + (2X + 1)H_2O \rightarrow R_3SiO\left[\begin{array}{c} Ar \\ | \\ Si-O \\ | \\ OSiR_3 \end{array}\right]_x SiR_3 + (4X + 2) HCl$$

where $R = C_2H_5$ and $Ar = C_6H_5$ or $p\text{-}ClC_6H_4$.

Similarly, Müller, Köhne and Sliwinski (43) cohydrolyzed mixtures of methyltrichlorosilane and trimethylchlorosilane to obtain the trimethylsiloxy-blocked series,

$$Me_3SiO\left[\begin{array}{c}Me\\ |\\ Si-O\\ |\\ OSiMe_3\end{array}-\right]_n SiMe_3 \qquad (n = 1—4) .$$

The analogous compounds, $Me_3SiOSiHOSiMe_3$ and $Me_3SiOSiHOSiHOSiMe_3$
    | |
  $OSiMe_3$ $OSiMe_3$ $OSiMe_3$
were prepared from trichlorosilane and trimethylchlorosilane.

IV. Silsesquioxanes

If an alkyl- or aryltrichlorosilane is treated, in bulk or in solution, with a considerable excess of water, an amorphous, infusible, insoluble product is usually formed with the approximate empirical composition, $(RSiO_{1.5})_x$ or $(ArSiO_{1.5})_x$. Insolubility has precluded estimation of the molecular weights of these products, and the amorphous appearance has discouraged crystallographic studies. These amorphous polymers are probably randomly cross-linked. Only in recent years have definable, low polymeric components of this composition been isolated and identified.

A crystalline silsesquioxane, octamethyloctasilsesquioxane, was first isolated by Scott (65) among the cracking products of mixed methylchlorosilane hydrolyzates. Its molecular weight was not established until some years later (75). This highly symmetrical octamer does not melt below 450° but is readily sublimable in a good vacuum at much lower temperatures.

Barry and Gilkey (11) discovered that low molecular weight polycyclic products, presumably $(C_2H_5SiO_{1.5})_8$, $(n–C_3H_7SiO_{1.5})_8$, and $(n–C_4H_9SiO_{1.5})_8$, were produced when initially oily hydrolysis products of the corresponding alkyltrichlorosilanes were heated with powdered alkali. These products and the cyclohexyl analogue were described more fully by Barry, Daudt, Domicone and Gilkey in a subsequent publication (10). Crystallographic study showed that the octamers were cubic, while the dodecamer, $(CH_3SiO_{1.5})_{12}$, also isolated during this work, showed hexagonal prismatic symmetry. Sprung and Guenther (75, 76) observed that the methyl- and ethyl-substituted octamers, formed in low yields when less than molar proportions of water were used in hydrolysis experiments, were sometimes accompanied by still smaller amounts of lower melting and more readily sublimable hexamers having the compositions $(CH_3SiO_{1.5})_6$ and $(C_2H_5SiO_{1.5})_6$. Crystallographic data were not reported, but the lower symmetry of these hexasilsesquioxanes is apparent from their structures and can be demonstrated by models. Based on analogies with the simpler and nearly planar trimeric and tetrameric dialkylsiloxanes, it is reasonable to suppose that the hexameric silsesquioxanes (*A*) are quite highly strained compared to the octameric silsesquioxanes (*B*). Strain energies have not as yet been calculated or measured.

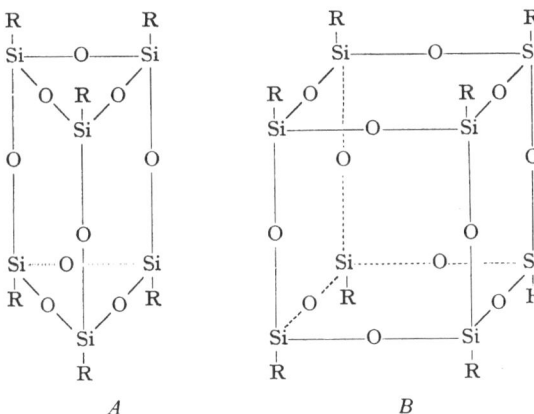

$$A \qquad\qquad B$$

Octa-*n*-amyloctasilsesquioxane, $(C_5H_{11}SiO_{1.5})_8$ is formed in relatively high yield on hydrolysis of *n*-amyltriethoxysilane in a high boiling ketone in the presence of a strongly alkaline catalyst (*78*). The bulky amyl groups are probably responsible both for the high yield and the low melting point of the product (this is the lowest melting crystalline silsesquioxane that has been reported). Octaphenyloctasilsesquioxane, $(C_6H_5SiO_{1.5})_8$, is obtained from phenyltriethoxysilane under somewhat similar conditions. However, the principal product formed was an amorphous high polymer that rearranged very readily to give the crystalline octamer when heated in inert solvents in the presence of traces of equilibration catalysts.

According to OLSSON (*53*) the octameric silsesquioxanes, $(RSiO_{1.5})_8$, where R is methyl, ethyl, *n*-propyl, isopropyl, *n*-butyl, and phenyl, are formed in relatively high yields (up to 44%) by hydrolysis of the corresponding alkyl- or aryltrichlorosilane under strongly acidic conditions.

Very recently MÜLLER, KÖHNE and SLIWINSKI (*44*) described the preparation of unsubstituted octasilsesquioxane $(HSiO_{1.5})_8$. This interesting prototype octamer was obtained in about 1% yield by treating trichlorosilane with 80% sulfuric acid in hexamethyldisiloxane solution. Like the methyl-, ethyl-, and phenyl-substituted octamers, it is a high melting solid, readily sublimable under vacuum.

These hexamers and octamers (and the single dodecamer so far reported) obviously represent members of polymer-homologous series that may, in time, be extended to include double-chain polysiloxanes of indefinitely high molecular weight. There can, of course, be no odd members of these series, since linkage of siloxanes in pairs is required by the simple mathematics and geometry of polysilsesquioxane structures. Presumably, too, the hypothetical dimers would involve entirely unrealistically strained Si–O–Si bonds, and should not be anticipated. But what about the tetramers?

This question was apparently answered a few years ago by Wiberg and Simmler (93), who isolated tetra-*tert*-butyltetrasilsesquioxane and tetra-*iso*-propyltetrasilsesquioxane. Wiberg and Simmler used a high dilution technique and relatively little water (1.5 moles of water per mole of alkyltrichlorosilane). The ether solvent and the hydrogen chloride formed on hydrolysis were removed under high vacuum, leaving the tetrasilsesquioxanes in high yield as readily sublimable solids. Under similar circumstances, methyl- and ethyltrichlorosilane produced only insoluble gels.

Apparently no further examples of tetrameric silsesquioxanes have yet been reported, nor have the Wiberg-Simmler experiments been repeated elsewhere to the knowledge of the present author.

Under somewhat similar conditions, the hydrolysis of trichlorosilane gives a crystalline, mica-like, high polymeric silsesquioxane (94). The structure of this interesting polymer is represented as a flat array of cyclohexasilsesquioxanes. On heating in a high vacuum to about 1000°, complete loss of hydrogen occurs, leaving a yellow-brown, crystalline, high polymeric material of empirical formula $(Si_2O_3)_x$.

V. Intermediate Hydrolysis Products from Tetrafunctional Monomers

The hydrolysis of silicon tetrachloride is extremely difficult to control, and although linear partial hydrolysis products such as hexachlorodisiloxane and octachlorotrisiloxane can be isolated in low yield under certain conditions (63) the propensity of this substance to go all the way to insoluble silica gels has discouraged extensive efforts in other directions.

On the other hand, if one or more of the reactive halogen atoms is first replaced by an alkoxyl group, it is possible to carry out more controllable hydrolyses. Thus ILER (*32*) studied the hydrolysis of di-*n*-butoxy-dichlorosilane in the presence of a tertiary amine and isolated a series of cyclic poly(di-*n*-butoxysiloxanes), $[n-(BuO)_2SiO]_x$ (x = 3–8). From *n*-butoxytrichlorosilane only insoluble gels were obtained, but cohydrolysis of the mono-and dibutoxy compounds gave liquid partial hydrolysis products that were postulated to consist of polycyclic siloxanes with *n*-butoxy and tri-*n*-butoxysiloxy substituents. Linear polyalkoxysiloxanes were prepared by MORGAN, OLDS and RAFFERTY (*41*) by treating silicon tetrachloride with methyl*iso*butyl carbinol or 2-ethylhexanol, followed by stepwise hydrolysis and condensation reactions. A partial hydrolysis scheme, obviously capable of endless variations, is shown below:

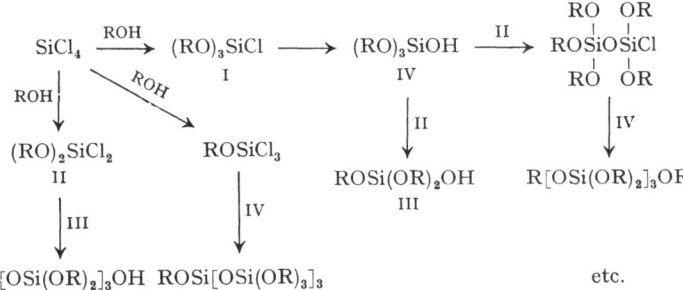

More recently OKAWARA, HOTTA and SHIMURA (*48*) hydrolyzed the more reactive dichlorodiethoxysilane in a benzene-pyridine mixture, isolated the corresponding linear dimer, tetramer, and pentamer, and obtained evidence for the presence of smaller amounts of the cyclic dimer, trimer and tetramer. The partial hydrolysis of tetramethoxy-, tetraethoxy-, tetra-*n*-butoxy- and tetra-*iso*-pentoxysilane in alcohol-water mixtures was studied by TAKATANI (*83*) and the following products isolated:

$$MeO\left[\begin{matrix}OMe\\|\\SiO\\|\\OMe\end{matrix}\right]_n Me \quad (n = 2\text{-}7), \qquad EtO\left[\begin{matrix}OEt\\|\\SiO\\|\\OEt\end{matrix}\right]_n Et \quad (n = 2\text{-}4),$$

$$BuO\left[\begin{matrix}OBu\\|\\SiO\\|\\OBu\end{matrix}\right]_n Bu \quad (n = 2,3), \qquad [(BuO)_2SiO]_4,$$

$$i\text{-}AmO\left[\begin{matrix}OAm\\|\\SiO\\|\\OAm\end{matrix}\right]_n Am \quad (n = 2,3), \qquad [(i\text{-}AmO)_2SiO]_4.$$

In the last two cases, higher boiling fractions obtained were postulated to have the following polycyclic structures:

$$
\begin{array}{ccccc}
& \underset{|}{\text{OR}} \;\; \underset{|}{\text{OR}} \;\; \underset{|}{\text{OR}} & & \underset{|}{\text{OR}} \;\; \underset{|}{\text{OR}} \;\; \underset{|}{\text{OR}} \;\; \underset{|}{\text{OR}} \\
& \text{RO}\overset{|}{\text{Si}}\text{-O-}\overset{|}{\text{Si}}\text{-O-}\overset{|}{\text{Si}}\text{OR} & & \text{RO}\overset{|}{\text{Si}}\text{-O-}\overset{|}{\text{Si}}\text{-O-}\overset{|}{\text{Si}}\text{-O-}\overset{|}{\text{Si}}\text{OR} \\
& \underset{|}{\overset{|}{\text{O}}} \;\; \underset{|}{\overset{|}{\text{O}}} \;\; \underset{|}{\overset{|}{\text{O}}} & \text{and} & \underset{|}{\overset{|}{\text{O}}} \;\; \underset{|}{\overset{|}{\text{O}}} \;\; \underset{|}{\overset{|}{\text{O}}} \;\; \underset{|}{\overset{|}{\text{O}}} \\
& \text{RO}\overset{|}{\text{Si}}\text{-O-}\overset{|}{\text{Si}}\text{-O-}\overset{|}{\text{Si}}\text{OR} & & \text{RO}\overset{|}{\text{Si}}\text{-O-}\overset{|}{\text{Si}}\text{-O-}\overset{|}{\text{Si}}\text{-O-}\overset{|}{\text{Si}}\text{OR} \\
& \overset{|}{\text{OR}} \;\; \overset{|}{\text{OR}} \;\; \overset{|}{\text{OR}} & & \overset{|}{\text{OR}} \;\; \overset{|}{\text{OR}} \;\; \overset{|}{\text{OR}} \;\; \overset{|}{\text{OR}}
\end{array}
$$

$R = n\text{-}C_4H_9$ or $iso\text{-}C_5H_{11}$

These structures are formally analogous with those postulated by SPRUNG and GUENTHER (75—78) for intermediates formed during the hydrolysis of trifunctional monomers.

Silicic Acids. The behavior of silicate ions in solution, the dependence of various properties on p_H, the nature of silica sols and gels, and the study of hydrated silicas constitute chapters in inorganic and colloid chemistry that go far beyond the scope of this review. Germane to the present subject, however, are certain observations on the formation of monosilicic acid and its stepwise polymerization.

The first convincing demonstration of the existence of monosilicic acid in solution is owing to WILLSTÄTTER and coworkers (36, 95). WILL-STÄTTER prepared monosilicic acid by careful treatment of silicon tetrachloride with moist silver oxide. The monomeric nature of the product was deduced from cryoscopic measurements. Its stability is very strongly p_H dependent, reaching a maximum at p_H 2—3. KRAUT's method was simply to adjust the p_H of an aqueous solution of $Na_2SiO_3 \cdot 6H_2O$ to the optimum. A further refinement of KRAUT's method is due to ALEXANDER (1) who treated hydrated sodium metasilicate with an aqueous slurry of a cation exchange resin at 0°, adjusted the p_H to 2.6 and filtered, obtaining a solution of silicic acid of D. P. $= 1.1$. ALEXANDER observed that polymerization occurred in solution most slowly at p_H 3.2 and more rapidly above or below this value. The mechanism of polymerization was different above and below the optimum — a third order rate being suggested at p_H 2 and a second order rate at higher p_H. The polymerization is very rapid above p_H 6, and is also accelerated by concentration, temperature and fluoride ion.

SCHWARZ and KNAUFF (64) approached the problem of the poly-silicic acids quite differently, by preparing linear partial hydrolysis products of methyl silicate,

$$
(\text{MeO})_3\text{SiO}\left[\begin{array}{c} \text{OMe} \\ | \\ \text{SiO} \\ | \\ \text{OMe} \end{array}\right]_n \text{Si(OMe)}_3 \qquad (n = 0,1,2)
$$

and cautiously hydrolyzing them. This should give the corresponding linear low-polymer silicic acids. Somewhat surprisingly, however, it was observed that Si–O–Si bonds appear to be broken under these conditions at least as readily as methoxy groups are saponified. Thus, the formation of monosilicic acid was observed. Its polymerization was again followed cryoscopically. The authors presented some evidence for ring formation in this system, yielding in one case a mixture consisting, on the average, of cyclic pentamers or hexamers. A theoretical explanation of the ease of formation of monosilicic acid from the linear methoxysiloxanes has been given by SCHOTT and FISCHER (62) based on the possibility of 3d electron shifts that confer a partial double bond character to the Si–OR bond and remove it from the Si–O–Si, thus weakening the latter. The acetoxysilanes would seem to be better suited to the purpose, because of their anhydride-like nature (expressed by SCHOTT and FISCHER as an incipient ionization with strengthening of the double bond character of the Si–O–Si); and, in truth, hydrolysis of hexaacetoxydisiloxane at pH 3.1 did produce disilicic acid, $(HO)_3SiOSi(OH)_3$. Like the monoacid, this polymerizes quite rapidly both below and above the optimum pH.

VI. Silanols

The hypothetical replacement of hydrogen by hydroxyl in silane, or in an alkyl or aryl silane, results in a series of silanols, formally analogous with the alcohol, vicinal glycol, hydrated carboxylic acid, and ortho-carbonic acid series of carbon chemistry. The silanols are in many ways both the most interesting and the most important silicone intermediates.

According to GRUBB and OSTHOFF (25) the physical properties of silanols are markedly similar to those of analogous carbinols. They are, for example, associated in the liquid state, presumably through hydrogen bonding. WEST (92) found, from infrared studies, that representative silanols are much more acidic than related carbinols, but only slightly less basic when judged by similar criteria.

A. Silanemonols

The monohydroxysilanes, usually obtained simply by hydrolysis of halogenosilanes, alkoxysilanes, or aryloxysilanes, range in properties from highly stable to extremely unstable liquids or solids. Most simple silanols lose water on distillation or melting, or sometimes at room temperature, to give the disiloxane. Silanol, H_3SiOH, has apparently not been isolated, nor have the simplest monoalkylsilanols. Dialkylsilanols and diaryl-silanols have been prepared; for example, diethylsilanol, $(C_2H_5)_2SiHOH$, by hydrolysis of tetraethyldisilazane, and ethylphenylsilanol and butyl-ethylsilanol by cautious hydrolysis of the corresponding chlorosilanes (67).

Silanols are, of course, greatly stabilized by the presence of bulky substituents. A methyl group is of insufficient bulk, and trimethylsilanol

30*

is consequently difficult to obtain free of the disiloxane. Sommer, Pietrusza and Whitmore (72) used trimethylfluorosilane for its preparation and obtained a 70% yield. This procedure has been followed rather extensively for the preparation of other unstable silanols. Weist (89), however, has pointed out certain advantages inherent in the use of trimethylchlorosilane in this preparation. Triethylsilanol is much more stable and is quite readily obtained free of the disiloxane. Higher trialkylsilanols are even more stable. Triisopropylsilanol is stable at the boiling point (196°) and tricyclohexylsilanol is apparently stable at the melting point, 177—8° (16). Triphenylsilanol is easily prepared, and not easily converted to the disiloxane, while tri-1-naphthylsilanol is stable to both alcoholic potash and to acids (21).

Large alkoxy groups also provide steric stabilization. For example, tri-tert-butoxysilanol, obtained by hydrolysis of sodium tri-tert-butoxysilanolate, can be vacuum distilled without decomposition (8).

B. Silanediols

Silanediol, $H_2Si(OH)_2$, was isolated by Stock and Somieski (80) but proved to be too unstable to keep more than a matter of minutes. The simplest alkylsilanediols, such as methylsilanediol, $CH_3SiH(OH)_2$ and ethylsilanediol, $C_2H_5SiH(OH)_2$, have apparently not been isolated. Dimethylsilanediol, $(CH_3)_2Si(OH)_2$, eluded investigation for many years because of its sensitivity to acid and alkaline reagents. It was isolated independently a few years ago by Hyde (30) and by Kantor (33). Diethylsilanediol, $(C_2H_5)_2Si(OH)_2$, is considerably more stable, yet many precautions must be observed during its preparation (15).

Dialkylsilanediols with bulky substituents are again increasingly stable. Di-n-propylsilanediol and di-n-butylsilanediol were prepared by George, Sommer and and Whitmore (20) by methods similar to that used for diethylsilanediol, while diisopropylsilanediol, diisobutylsilanediol, and dicyclohexylsilanediol are readily prepared from the corresponding difluorosilanes (16). The last of these was, however, obtained from dicyclohexyldichlorosilane during the early Kipping researches (14). Di-tert-butylsilanediol is very difficult to dehydrate, and when a single tert-butyl group is present, as in tert-butylmethylsilanediol, tert-butylphenylsilanediol, and tert-butylhexadecylsilanediol, the products are still very resistant to dehydration (73). Tertiary alkoxy groups exert somewhat lower steric effects. Thus, di-tert-butoxysilanediol is stable only when quite pure (40).

The alkylarylsilanediols are quite stable as a class, and several were prepared during the early Kipping researches (14, 57). Recently Shostakovskii et al. have described the preparation of alkylphenyl- and alkylnaphthylsilanediols by hydrolysis of the corresponding diacetoxysilanes

(*68, 69*). Alkenylarylsilanediols are also readily prepared by hydrolysis of the diacetoxysilane. In this way FRISCH, GOODWIN and SCOTT (*19*) prepared phenylvinyl-, benzylvinyl-, *p*-chlorophenylvinyl-, allylphenyl-, and allylbenzylsilanediol.

Diarylsilanediols are generally quite easy to prepare by any of the methods mentioned above, and are the most stable representatives of their class.

C. Polysiloxanediols

Disiloxanediols are sometimes obtained inadvertantly when the silanediol is desired, or appear as by-products of the latter. Synthesis of the lower members is often quite difficult, and careful control of pH and temperature was found necessary by LUCAS and MARTIN (*38*) to prepare tetramethyldisiloxanediol. FRISCH, GOODWIN and SCOTT (*19*) obtained a number of divinyl- and diallyldisiloxanediols under conditions similar to those used for the preparation of the related silanediols. ROBINSON and KIPPING (*57*) early obtained relatively stable disiloxanediols from silanediols by means of heat or acids.

The poly-(diphenylsiloxane)-diols, $HO(Si\emptyset_2O)_nH$ (n = 2,3,4) were obtained by KIPPING and ROBISON (*35*) by treating diphenylsilanediol with ammonium hydroxide; and by BURKHARD (*13*) by hydrolyzing the corresponding α, ω-dichlorides in *tert*-amyl alcohol-toluene solution. Tetra-*p*-chlorophenyldisiloxanediol and hexa-*p*-chlorophenyl-1,5-tri-siloxanediol were recently isolated by SCHOTT and BERGE (*61*).

D. Silanetriols

Silanetriols eluded all attempts at their isolation until a few years ago when TYLER (*88*) prepared phenylsilanetriol from phenyltrimethoxysilane and dilute acetic acid at about 10°. More recently TAKIGUCHI (*82*) obtained this substance from phenyltrichlorosilane in the presence of aniline. ANDRIANOV, ZHDANOV and MORGANOVA (*7*) obtained dichlorophenylsilanetriol by hydrolyzing the triacetoxysilane. The interesting tetrol, bis-(dichlorophenyl)-disiloxanetetrol, was similarly obtained from bis (dichlorophenyl)-tetraacetoxydisiloxane.

E. Silanetetrol

Silanetetrol (monosilicic acid) has been prepared only in solution and is stable only within a very narrow pH range (see section V).

F. Reactions of Silanols

Considering their scientific interest and technological importance, the reactivities of silanols have been given scant attention. Most of the reactions by which silanols are prepared are found to be reversible under the proper conditions. For example, SOMMER, PIETRUSZA and WHITMORE (72) converted triethylsilanol to chlorotriethylsilane, to triethylsiloxyl acetate, to bis-(triethylsiloxyl) sulfate and to sodium triethylsilanolate. Alkali silanolates can be obtained from certain silanols by treatment either with the alkali amide, or with concentrated alkali metal hydroxide (29, 31, 86, 87). SAUER (59) prepared trimethylsiloxymagnesium iodide from trimethylsilanol and methylmagnesium iodide. Recently, MULLER, DATHE and HEINRICH (42) observed that certain silanols can be methylate with diazomethane in alcohol solution.

BOEHM and SCHNEIDER (12) have observed that many of these reactions can be made to occur with the free silanol groups on the surface of finely divided, amorphous silica gel.

The reaction of a silanol with an alcohol to give an alkoxysilane is quite common in practice, but has received relatively little study. A few years ago (23), GILMAN and MILLER pointed out that the silanol-alcohol reactions must be taken into consideration in order to account for the observed stoichiometry in the titration of silanols with KARL FISCHER reagent. Since a mole of silanol titrates as though it were a mole of water, it seemed that the silanol must react with the large excess of methanol present to give methoxysilane and water. It was noted, further, that the equilibrium was attained slowly when bulky substituents were present in the silanol. The stoichiometry of the KARL FISCHER titration would be upset if silanol condensed to a significant extent to give disiloxane, since one mole of silanol then produces only one-half mole of water.

GRUBB considered this problem in detail and concluded (24) that under these conditions the reaction of silanols with methanol is rapid compared to self-condensation. Even under milder conditions the self-condensation of silanols is not complete when much alcohol is present, but an equilibrium is established in which most of the silanol is present as alkoxysilane. GRUBB discussed the equilibria involved:

$$-\text{SiOH} + \text{MeOH} \underset{k_2}{\overset{k_1}{\rightleftarrows}} -\text{SiOMe} + \text{H}_2\text{O} \tag{1}$$

$$-\text{SiOH} + \text{SiOMe} \underset{k_4}{\overset{k_3}{\rightleftarrows}} -\text{SiOSi}- + \text{MeOH} \tag{2}$$

$$2-\text{SiOH} \underset{k_6}{\overset{k_5}{\rightleftarrows}} -\text{SiOSi}- + \text{H}_2\text{O} \tag{3}$$

and deduced the constant,

$$K_4 = K_1 K_2 = \frac{k_1}{k_2} \cdot \frac{k_3}{k_4} = \frac{(-\text{SiOMe})^2(\text{H}_2\text{O})}{(-\text{SiOSi}-)} .$$

The validity of this expression indicated that the reaction between silanol and methanol was essentially complete. K_4 decreased in going from MeOH to EtOH to n-PrOH (i. e., relatively more –SiOSi– is present at equilibrium) and the equilibrium was reached more slowly if bulky organic groups were attached to the silicon. A mechanism consistent with the observed facts is "back-side displacement" of the alkoxy group by silanol or by silanolate ion.

Acid catalyzed:

$$R_3SiO: + \overset{R_3}{\underset{\underset{\overset{..}{H}\oplus}{H}}{SiOMe}} \rightarrow R_3SiOSiR_3 + MeO\overset{\oplus}{H_2} \tag{4}$$

Base catalyzed:

$$R_3Si\overset{\ominus}{O}: + \overset{R_3}{SiOMe} \rightarrow R_3SiOSiR_3 + Me\overset{\ominus}{O} \tag{5}$$

Steric factors should be of considerable importance in either case.

Sprung and Guenther (79) studied the acid or base catalyzed reaction of silanols with n-octanol under non-equilibrium conditions, and found that self-condensation of the silanol and alcoholysis of the resulting disiloxane were always competing reactions. In the presence of acids, the observed rate sequence was $Et_3SiOH > Me_3SiOH \gg \emptyset_3SiOH$; while in the presence of base the sequence was $Me_3SiOH > \emptyset_3SiOH \gg Et_3SiOH$. The steric influence of the ethyl substituents was more pronounced when the catalyst was a base rather than an acid. The normal inductive effects of both electron attracting and electron releasing groups can be observed in either case. Inductive effects were also observed by Schott and Berge (60) in the thermal self-condensation of silanols, in the order $p\text{–}ClC_6H_4 > p\text{–}BrC_6H_4 > C_6H_5 > C_6H_5CH_2$; while resonance effects, in the order $p\text{–}BrC_6H_4 > p\text{–}ClC_6H_4 > C_6H_5$ were considered to explain the dephenylation of arylsilanediols observed to occur at the high temperatures involved.

VII. Mechanism of Solvolysis of Reactive Silanes

A. The Silicon-Hydrogen Bond

Highly reactive halogen-substituted silanes were understandably avoided in earlier mechanism studies, and attention was devoted instead to the hydrolytic and solvolytic cleavage of the more stable Si–H bond. The reaction, most simply represented by the expression, $\geqslant SiH + H_2O \rightarrow \geqslant SiOH + H_2$, is strongly base catalyzed. However, as Price (55) pointed out, the stoichiometry is represented better by the equation: $\geqslant SiH + ROH + OH^- \rightarrow \geqslant SiOH + OR^- + H_2$. Price found that the rate of hydrogen evolution is given by the expression: $k_1 t = \ln [V_F / V_F - V]$

where V_F and V represent the volumes of gas evolved at infinite time and at time, t, and k, is the apparent first order rate constant. The rate expression was more precisely written in the form: $\frac{-dx}{dt} = k_2 [x][OH^-]$, where x = silane, but the rate was also apparently proportional to the concentration of the hydroxylated solvent employed. It was postulated that the reaction involves nucleophylic attack of hydroxyl ion on silicon with ejection of a hydride ion; and that this is then immediately attacked by a proton from the hydroxylated solvent. Steric effects were found to be important; bulky groups on either the silicon atom of the silane or the oxygen atom of the solvent greatly reduced the rate.

GILMAN and DUNN (22) observed that the hydrolysis of triaryl silanes in piperidine was pseudo first order, on observation consistent with PRICE's mechanism. The influence of electron attracting and releasing groups in the phenyl groups and the effect of their position is evident from the following order of reactivity ($k_1 \times 10^4$): p–Cl,16.9; none, 3.22; m–CH$_3$, 2.75; p–CH$_3$, 1.08; p–OCH$_3$, 0.89; m–N(CH$_3$)$_2$, 0.75; p–N(CH$_3$)$_2$, 0.21. Electron releasing groups obviously exert a strong retarding effect, especially when in the p-position.

WEST (91), amplifying on these results, argued that since the solvolysis is bimolecular it must proceed either through a normal S_N2 bimolecular displacement or involve a rather stable pentacovalent intermediate. Both mechanisms, WEST believes, must involve a 5-coordinate transition state, and therefore may really be thought of as equivalent. WEST found that silacyclopentane was 13 times as reactive as diethylmethylsilane and 200 times as reactive as silacyclohexane (which could be construed as evidence for I-strain in silacyclopentane). Since this order of reactivity is the same as that found in carbocyclic compounds, it was concluded that similar considerations of energy and entropy of reaction are encountered, a possibility that had also been advanced by PRICE.

Two mechanisms that could account for all the observed facts were proposed by KAPLAN and WILZBACH (34).

A. $R_3SiH + OH^- \xrightarrow{\text{slow}} \left[R_3Si{<}^{OH}_{H} \right]^-$

$\left[R_3Si{<}^{OH}_{H} \right]^- + HS \xrightarrow{\text{fast}} R_3SiOH + H_2 + S^-$

B. $R_3SiH + OH^- + HS \longrightarrow R_3SiOH + H_2 + S^-$

Mechanism B could involve either a single termolecular step; rapid reaction of solvent with an intermediate, slowly formed pentacovalent complex; or a "concerted" attack of OH^- and solvent on the silane. KAPLAN and WILZBACH found that substitution of deuterium or tritium in either the silane or the solvent lowered the reaction rate. This, they felt, means that the rate determining step is the rupture of the Si–H bond; and that, in the transition state, the H atom from the silane is already strongly bonded to the H atom from the solvent, as for example in

$$
\left[
\begin{array}{c}
R_3Si\text{---}H\text{---}H\text{--}S \\
| \\
| \\
OH
\end{array}
\right]^- .
$$

The acid catalyzed solvolysis of the Si–H bond was studied by BAINES and EABORN (9). In 95% ethanol, 1.43 molar in HCl, the following first order rate constants were observed at 34.9° ($k_1 \times 10^3$): $p\text{-}ClC_6H_4SiMe_2H$, 2,58; $C_6H_5SiMe_2H$, 2.13; $p\text{-}MeC_6H_4SiMe_2H$, 1.82; $(p\text{-}ClC_6H_4)_3SiH$, 1.50; Et_3SiH, 1.19; $n\text{-}Pr_3SiH$, 0.75; $n\text{-}Bu_3SiH$, 0.70; $(C_6H_5)_3SiH$, 0.40; $(t\text{-}Bu)_3SiH$, 0.20; $(i\text{-}Pr)_3SiH$, 0.07. At 45.0°, the first order constants ($k_1 \times 10^3$) were in the same order: $C_6H_5SiMe_2H$, 5.52; $p\text{-}MeC_6H_4SiMe_2H$, 5.02; Et_3SiH, 3.03; $n\text{-}Pr_3SiH$, 2.07; $n\text{-}Bu_3SiH$, 1.95. Both steric and electrical effects can be detected. Thus, the effect of electron releasing groups in the p-position of a phenyl group is clearly to retard the reaction. BAINES and EABORN propose that the acid catalyzed reaction involves electrophylic attack of an oxonium ion on the Si–H bond. Since the reaction is retarded by the presence of groups that can cause an increase in the electron density around the silicon atom, it is argued that the transition state must be one of relatively high negative charge that facilitates nucleophylic attack by a solvent molecule. A two-step mechanism was proposed:

$$>SiH + H_2O \xrightarrow[\text{fast}]{} \overset{\ominus}{Si}H\text{--}\overset{\oplus}{H}OH$$

$$>\overset{\ominus}{Si}H\text{--}\overset{\oplus}{H}OH + H_3\overset{\oplus}{O} \xrightarrow{\text{slow}} >Si\overset{\oplus}{O}H_2 + H_2 + H_2O .$$

It should be noted that this transition state does not differ too greatly from that postulated for the alkaline catalyzed reaction by KAPLAN and WILZBACH. The latter, however, is formed by the reaction of a solvent molecule with a pentacovalent negative ion formed in a rate controlling step; whereas in the other instance, the rate controlling step is the attack of H_3O^+ on the neutral transition complex. Further consideration and reconciliation of these views will obviously be of interest.

B. The Silicon-Halogen Bond

The formation of a pentacovalent silicon intermediate has been advanced frequently as the explanation of the ease of solvolysis of the silicon-halogen bond. ROCHOW, for example, (58) places considerable emphasis on the large size of silicon compared to carbon, and the potentiality for expansion of the valence shell of silicon from 8 to 10 is considered to account for the greater reactivity of substituted silanes in general. This hypothesis was evoked also by Swain, Esteve and Jones (81) who studied the hydrolysis of fluorotriphenylsilane in 50% aqueous acetone, and observed that it was very sensitive to alkali, retarded by electron releasing groups in the p-position, and only weakly sensitive to high concentrations of water or to neutral salts, all in contradistinction to the behavior of the carbon-fluorine bond under similar conditions. They concluded that a siliconium ion could not be involved, but that an intermediate was indicated having a lesser positive charge on silicon than the silane itself, and suggested that this could be a pentacovalent silicon intermediate. However, Hughes (27) suggested that a formal $S_N 2$ reaction would explain all the existing data without recourse to a pentacovalent intermediate. This view has received support more recently from Allen, Charlton. Eaborn and Modena (2) who studied the hydrolysis and alcoholysis of chlorotriisopropylsilane in various alcohols and postulate a simple $S_N 2$ mechanism; namely,

$$R'OH + R_3SiCl \rightarrow \left[\begin{array}{c} R \quad R \\ \diagdown \diagup \\ R'O\text{---}Si\text{---}Cl \\ | \quad | \\ H \quad R \end{array} \right] \rightarrow R'OSiR_3 + HCl .$$

Referring to the effect of electron releasing groups, they suggest that these depress the rate of solvolysis merely because in this $S_N 2$ reaction bond-forming is more important than bond-breaking, and electropositive substituents would be expected adversely to affect the bond-forming requirements.

Because triphenylchlorosilane is much more reactive toward solvolysis than triisopropylchlorosilane (whereas on steric grounds they should be about equal) Allen and Modena (3) were further convinced that the presence on silicon of easily polarizable groups strongly facilitates formation of the transition state. Based on a study of solvolysis in dioxane at 25.1° they derived the following rate expression:

$$\text{Rate} = k_1(H_2O)[R_3SiCl] + k_3[R_3SiCl][H_2O][Cl^-]$$

Where $k_1(H_2O)$ is the first order rate constant for the uncatalyzed reaction at a given water concentration, $k_3 = k_2/[H_2O]$, $k_2 = k_c/[Cl^-]$, $k_c = k_1$ (catalyzed) $+ k_1$ (uncatalyzed), and $[Cl^-]$ is the concentration of chloride ion from neutral salts. They again oppose the Swain mechanism on the basis a) that the unimolecular dissociation of the pentacovalent complex should be subject to electrophylic catalysis, b) that the steric effects are too great, c) that symmetrical chloride exchange is not fast kinetically, and d) that catalysis by chloride ion is again difficult to account for; and again propose an S_N2 type reaction path:

$$R'OH----B + R_3SiCl---- \left[\begin{array}{c} R \diagdown\ \diagup R \\ R'O---\underset{\underset{\displaystyle B---H}{|}}{\overset{\overset{\displaystyle R}{|}}{Si}}---Cl \end{array} \right]$$

$$\downarrow$$

$$R'OSiR_3 + HB^+ + Cl^-$$

Conductometric measurements were used in a recent study of the hydrolysis of alkyldi- and trichlorosilanes in polar solvents by SHAFFER and FLANIGEN (66). A first order dependence on water concentration and a second order dependence on chlorosilane was observed for the hydrolysis of trichlorosilanes. The dichlorosilanes evinced a highly anamolous negative order. For the trichlorosilanes the data are consistent with a mechanism that assumes the following steps:

1. $RSiCl_3 + H_2O \cdot HCl \cdot S(S = solvent) \rightleftharpoons RSiCl_2OH + 2HCl \cdot S$

2. $RSiCl_2OH + RSiCl_3 + S \longrightarrow (RSiCl_2)_2O + HCl \cdot S$

The rate expression is then

$$Rate = \frac{k_2[H_2O][RSiCl_3]^2}{\dfrac{[HCl]}{K_1} + \dfrac{k_2}{k_1}[RSiCl_3]} \cdot$$

These data are apparently in overall agreement with the mechanisms of SCHUMB and STEVENS (63) and of SWAIN, ESTEVE and JONES (81). The ALLEN and MODENA data were apparently not available at this time and were not discussed in this paper.

C. The Si-OR Bond

The kinetics of alkoxysilane hydrolysis have not yet been resolved successfully. REUTHER (56) published a preliminary report a few years ago on conductometric measurements on some alkoxysilanes, but kinetic expressions were not devised. SHAFFER and FLANIGEN (66) also found that conductometric data obtained on hydrolysis of alkoxysilanes were difficult to interpret.

References

1. ALEXANDER, G. B.: J. Am. Chem. Soc. 75, 2887 (1953); 76, 2094 (1954).
2. ALLEN, A. D., J. C. CHARLTON, C. EABORN and G. MODENA: J. Chem. Soc. 1957, 3668.
3. — and G. MODENA: J. Chem. Soc. 1957, 3671.

4. Andrianov, K. A.: J. Gen. Chem. U.S.S.R. **8**, 1255 (1938); Chem. Abs. **33**, 4193 (1939).

5. — M. Ya Levshuk, S. A. Golubtsov and T. A. Krasovskaya: Zhur. Obshchei Khim. **28**, 333 (1958); Chem. Abs. **52**, 13663 (1958).

6. — and L. I. Makarova: Izvest. Akad. Nauk, Otdel. Khim. Nauk **1959**, 450; Chem. Abs. **53**, 21622 (1959).

7. — A. A. Zhdanov and E. F. Morgunova: Zhur. Obshchei Khim. **27**, 156 (1957). Chem. Abs. **51**, 12845 (1957).

8. Backer, H. J., and H. A. Klasens: Rec. trav. chim. **61**, 511 (1942).

9. Baines, J. E., and C. Eaborn: J. Chem. Soc. **1956**, 1436.

10. Barry, H. J., W. H. Daudt, J. J. Domicone and J. W. Gilkey: J. Am. Chem. Soc. **77**, 4248 (1955).

11. — and J. W. Gilkey: U. S. Patent 2, 465, 188, March 22, 1949.

12. Boehm, H. P., and M. Schneider: Z. anorg. allgem. Chem. **301**, 326 (1959).

13. Burkhard, C. A.: J. Am. Chem. Soc. **67**, 2173 (1945).

14. Cusa, N. W., and F. S. Kipping: J. Chem. Soc. **1932**, 2205.

15. Di Giorgio, P. A., L. H. Sommer and F. C. Whitmore: J. Am. Chem. Soc. **68**, 344 (1946).

16. Eaborn, C.: J. Chem. Soc. **1952**, 2840.

17. Fletcher, H. J., and M. J. Hunter: J. Am. Chem. Soc. **71**, 2918 (1949); Cf. K. A. Andrianov: J. Gen. Chem. U.S.S.R. **16**, 633 (1946).

18. — — J. Am. Chem. Soc. **71**, 2922 (1949).

19. Frisch, K. C., P. A. Goodwin and R. E. Scott: J. Am. Chem. Soc. **74**, 4584 (1952).

20. George, P. D., L. H. Sommer and F. C. Whitmore: J. Am. Chem. Soc. **75**, 1585 (1953).

21. Gilman, H., and C. G. Brannen: J. Am. Chem. Soc. **73**, 4640 (1951).

22. — and G. E. Dunn: J. Am. Chem. Soc. **73**, 3404 (1951).

23. — and L. S. Miller: J. Am. Chem. Soc. **73**, 2367 (1951).

24. Grubb, W. T.: J. Am. Chem. Soc. **76**, 3408 (1954).

25. — and R. C. Osthoff: J. Am. Chem. Soc. **75**, 2230 (1953).

26. Hardy, D. V. N., and N. J. L. Megson: "The Chemistry of Silicone Polymers" Quart. Rev. (London) **2**, 25 (1948).

27. Hughes, E. D.: Quart. Rev. (London) **5**, 245 (1951).

28. Hunter, M. J., J. F. Hyde, E. L. Warrick and H. J. Fletcher: J. Am. Chem. Soc. **68**, 667 (1946).

29. Hyde, J. F.: U.S. Patent 2, 567, 110, September 4, 1951.

30. — J. Am. Chem. Soc. **75**, 2166 (1953).

31. — O. K. Johannson, W. H. Daudt, R. F. Fleming, H. B. Laudenslager and M. P. Roche: J. Am. Chem. Soc. **75**, 5615 (1953).

32. Iler, R. K.: J. Ind. Eng. Chem. **39**, 1384 (1947); U.S. Patent 2, 499, 865 (March 7, 1950).

33. Kantor, S. W.: J. Am. Chem. Soc. **75**, 2712 (1953).

34. Kaplan, L., and K. E. Wilzbach: J. Am. Chem. Soc. **77**, 1297 (1955).

35. Kipping, F. S., and R. Robison: J. Chem. Soc. **1914**, 484.

36. Kraut, H.: Ber. dtsch. chem. Ges. **64**, 1709 (1931).

37. Lasocki, Z.: Roczniki Chem. **31**, 305, 837 (1957); Chem. Abs. **51**, 16282 (1957); **52**, 10005 (1958).

38. Lucas, G. R., and R. W. Martin: J. Am. Chem. Soc. **74**, 5225 (1952).

39. Matsui, M.: J. Sci. Research Inst. (Tokyo) **51**, 225 (1957); Chem. Abs. **52**, 12445 (1958).

40. MINER, C. S. Jr., L. A. BRYAN, R. P. HOLYSZ and G. W. PEDLOW Jr.: J. Ind. Eng. Chem. **39**, 1368 (1947).
41. MORGAN, C. R., W. F. OLDS and A. L. RAFFERTY: J. Am. Chem. Soc. **73**, 5193 (1951).
42. MÜLLER, R., C. DATHE and L. HEINRICH: J. prakt. Chem. [4], **9**, 24 (1959).
43. — R. KÖHNE and S. SLIWINSKI: J. prakt. Chem. [4], **9**, 63 (1959).
44. — — — J. prakt. Chem. [4], **9**, 71 (1959).
45. OKAWARA, R.: Bull. Chem. Soc. Japan **27**, 428 (1954); Chem. Abs. **50**, 162 (1956).
46. — Bull. Chem. Soc. Japan **31**, 154 (1958).
47. — T. ANDO and K. AYAMA: Technol. Repts. Osaka Univ. **8**, 171 (1958); Chem. Abs. **53**, 12160 (1959).
48. — S. HOTTA and T. SHIMURA: Bull. Chem. Soc. Japan **28**, 541 (1955); Chem. Abs. **50**, 11939 (1956).
49. — and S. IMAEDA: Bull. Chem. Soc. Japan **31**, 194 (1958); Chem. Abs. **52**, 15422 (1958).
50. — and I. ISHIMARU: Bull. Chem. Soc. Japan **27**, 582 (1954); Chem. Abs. **50**, 162 (1956).
51. — G. MINAMI and Z. OKU: Bull. Chem. Soc. Japan **31**, 22 (1958); Chem. Abs. **52**, 15422 (1958).
52. — and M. SAKIYAMA: Technol Repts. Osaka Univ. **7**, 459 (1957); Chem. Abs. **53**, 3039 (1959).
53. OLSSON, K.: Arkiv. Kemi **13**, 367 (1958); Chem. Abs. **53**, 17887 (1959).
54. PATNODE, W. I., and D. F. WILCOCK: J. Am. Chem. Soc. **68**, 358 (1946).
55. PRICE, F. P.: J. Am. Chem. Soc. **69**, 2600 (1947).
56. REUTHER, H.: Z. anorg. allgem. Chem. **272**, 122 (1953).
57. ROBINSON, R., and F. S. KIPPING: J. Chem. Soc. **1912**, 2142, 2158, 2164.
58. ROCHOW, E. G.: "The Chemistry of the Silicones", John Wiley and Sons, 2nd Edition 1951.
59. SAUER, R. O.: J. Am. Chem. Soc. **66**, 1707 (1944).
60. SCHOTT, G., and H. BERGE:. Z. anorg. allgem. Chem. **297**, 32 (1958).
61. — — Z. anorg. allgem. Chem. **297**, 44 (1958).
62. — and E. FISCHER: Z. anorg. allgem. Chem. **301**, 179 (1959).
63. SCHUMB, W. C., and A. J. STEVENS: J. Am. Chem. Soc. **69**, 726 (1947); **72**, 3178 (1950).
64. SCHWARZ, R., and K. G. KNAUFF: Z. anorg. allgem. Chem. **275**, 176 (1954).
65. SCOTT, D. W.: J. Am. Chem. Soc. **68**, 356 (1946).
66. SHAFFER, L. H., and E. M. FLANIGEN: J. Phys. Chem. **61**, 1591, 1595 (1957).
67. SHOSTAKOVSKII, M. F., D. A. KOCHKIN and V. M. ROGOV: Izvest. Acad. Nauk, Otdel Acad. Nauk, **1956**, 1062; Chem. Abs. **51**, 4983 (1957).
68. — — KH. I. KONDRATEV and V. M. ROGOV: Zhur. Obshchei Khim. **26**, 3344 (1956); Chem. Abs. **51**, 9514 (1957).
69. — and KH. I. KONDRATEV: Izvest. Acad. Nauk, Otdel. Khim. Nauk **1956**, 967; Chem. Abs. **51**, 4983 (1957).
70. SOKOLOV, N. N.: Zhur. Obshchei. Khim. **29**, 258 (1959); Chem. Abs. **53**, 21622 (1959).
71. — and K. A. ANDRIANOV: Izvest. Akad. Nauk, Otdel. Khim. Nauk, **1957**, 806; Chem. Abs. **52**, 3668 (1958).
72. SOMMER, L. H., E. W. PIETRUSZA and F. C. WHITMORE: J. Am. Chem. Soc. **68**, 2282 (1946).
73. — and L. J. TYLER: J. Am. Chem. Soc. **76**, 1030 (1954).

74. Sprung, M. M., and F. O. Guenther: J. Am. Chem. Soc. 77, 4173 (1955); See also M. Matsui: J. Sci. Res. Inst. (Tokyo) 51, 225 (1957); Chem. Abs. 52, 12445 (1958).
75. — — J. Am. Chem. Soc. 77, 3990 (1955).
76. — — J. Am. Chem. Soc. 77, 3996 (1955).
77. — — J. Am. Chem. Soc. 77, 6045 (1955).
78. — — J. Polym. Sci. 28, 17 (1958).
79. — — J. Org. Chem. (In Press.)
80. Stock, A., and C. Somieski: Ber. 52, 695 (1919).
81. Swain, C. G., R. M. Esteve and R. H. Jones: J. Am. Chem. Soc. 71, 965 (1949).
82. Takiguchi, T.: J. Am. Chem. Soc. 81, 2359 (1959).
83. Takatani, T.: Nippon Kagaku Zasshi 76, 7, 679 (1955); Chem. Abs. 51, 17724 (1957).
84. Tamborski, C., and H. W. Post: J. Org. Chem. 17, 1400 (1952).
85. Tanaka, T., A. Tasaka, R. Okawara and T. Watase: Technol. Repts. Osaka Univ. 7, 193 (1957); Chem. Abs. 52, 9950 (1958).
86. Tatlock, W. S., and E. G. Rochow: J. Am. Chem. Soc. 72, 528 (1950).
87. — — J. Org. Chem. 17, 1555 (1952).
88. Tyler, L. J.: J. Am. Chem. Soc. 77, 771 (1955).
89. Weist, M.: Chem. Tech. 5, 303 (1953).
90. West, R.: J. Am. Chem. Soc. 75, 1002 (1953).
91. — J. Am. Chem. Soc. 76, 6015 (1954).
92. — and R. H. Baney: J. Am. Chem. Soc. 81, 6145 (1959).
93. Wiberg, E., and W. Simmler: Z. anorg. allgem. Chem. 282, 330 (1955).
94. — — Z. anorg. allgem. Chem. 283, 401 (1956).
95. Willstatter, R., H. Kraut and K. Lobinger: Ber. 58, 2462 (1925); 61, 2280 (1928); 62, 2027 (1929).

Elektronen- und Ionenprozesse in Ionenkristallen mit Berücksichtigung photochemischer Prozesse

Von Professor Dr. Ostap Stasiw

Institut für Kristallphysik Berlin-Adlershof

Struktur und Eigenschaften der Materie in Einzeldarstellungen, Bd. XXII

Mit 107 Abbildungen. VIII, 307 Seiten Gr.-8°. 1959

Ganzleinen DM 66,—

Statistik von Störstellen in Ionenkristallen. Fehlordnungsenergie. Platzwechselvorgänge, Diffusion und Ionenleitung. Das Absorptionsspektrum des idealen Ionengitters. Das Absorptionsspektrum von Ionengittern mit stöchiometrischem Überschuß der Kationen- oder Anionenkomponente. Absorptionsspektren von Ionengittern mit Fremdzusätzen. Elektronische Störstellentheorie. Halbleiterprozesse. Lichtelektrische Leitung. Photochemische Prozesse in reinen Ionengittern, in Ionengittern mit Zusätzen, in mechanisch verformten Kristallen. — Störstellen und Kernresonanz. Anwendung der adiabatischen Näherung auf Kristalle mit Störstellen

Aus den Besprechungen

In den letzten 30 Jahren sind so zahlreiche Veröffentlichungen über diesen Problemkreis erschienen, daß selbst für einen Fachkollegen auf dem Gebiet der Festkörperphysik das Sammeln und kritische Überdenken der zahlreichen experimentellen Ergebnisse und theoretischen Ansätze auf diesem Teilgebiet nur mit Mühe möglich ist, geschweige denn für einen außenstehenden Kollegen. Es ist daher zu begrüßen, daß gerade O. Stasiw, als besonders guter Kenner dieses Fachgebietes bekannt durch seine zahlreichen Beiträge, sich der Mühe unterzog, die Vielzahl der Einzelbeobachtungen und Mitteilungen in einer gut übersichtlichen Weise kritisch niederzuschreiben. Neben der klaren und verständlichen Darstellung verdienen zwei Merkmale derselben besonders hervorgehoben zu werden: Erstens, daß der Leser neben dem Bericht über feststehende Tatsachen zahlreiche Anregungen zur weiteren Arbeit findet, und zweitens, daß die unumgängliche physikalisch-mathematische Behandlung der Absorptionsprozesse in das letzte Kapitel verlegt wurde, um dem theoretisch nicht genügend vorgebildeten Leser das Studium dieser Vorgänge nicht unnötig zu erschweren.

Kolloid-Zeitschrift

SPRINGER-VERLAG · BERLIN · GÖTTINGEN · HEIDELBERG